Systems and Technologies for the Treatment of Non-Stockpile Chemical Warfare Materiel

Committee on Review and Evaluation of the Army
Non-Stockpile Chemical Materiel Disposal Program

Board on Army Science and Technology

Division on Engineering and Physical Sciences

National Research Council

NATIONAL ACADEMY PRESS
Washington, D.C.

NATIONAL ACADEMY PRESS   2101 Constitution Avenue, N.W.   Washington, DC 20418

NOTICE: The project that is the subject of this report was approved by the Governing Board of the National Research Council, whose members are drawn from the councils of the National Academy of Sciences, the National Academy of Engineering, and the Institute of Medicine. The members of the committee responsible for the report were chosen for their special competences and with regard for appropriate balance.

This is a report of work supported by Contract DAAD19-01-C-008 between the U.S. Army and the National Academy of Sciences. Any opinions, findings, conclusions, or recommendations expressed in this publication are those of the author(s) and do not necessarily reflect the views of the organizations or agencies that provided support for the project.

International Standard Book Number 0-309-08452-0

Limited copies are available from:

Board on Army Science and Technology
National Research Council
2101 Constitution Avenue, N.W.
Washington, DC 20418
(202) 334-3118

Additional copies are available for sale from:

National Academy Press
Box 285
2101 Constitution Ave., N.W.
Washington, DC 20055
(800) 624-6242 or (202) 334-3313
(in the Washington metropolitan area)
http://www.nap.edu

Copyright 2002 by the National Academy of Sciences. All rights reserved.

Printed in the United States of America

THE NATIONAL ACADEMIES

National Academy of Sciences
National Academy of Engineering
Institute of Medicine
National Research Council

The **National Academy of Sciences** is a private, nonprofit, self-perpetuating society of distinguished scholars engaged in scientific and engineering research, dedicated to the furtherance of science and technology and to their use for the general welfare. Upon the authority of the charter granted to it by the Congress in 1863, the Academy has a mandate that requires it to advise the federal government on scientific and technical matters. Dr. Bruce M. Alberts is president of the National Academy of Sciences.

The **National Academy of Engineering** was established in 1964, under the charter of the National Academy of Sciences, as a parallel organization of outstanding engineers. It is autonomous in its administration and in the selection of its members, sharing with the National Academy of Sciences the responsibility for advising the federal government. The National Academy of Engineering also sponsors engineering programs aimed at meeting national needs, encourages education and research, and recognizes the superior achievements of engineers. Dr. Wm. A. Wulf is president of the National Academy of Engineering.

The **Institute of Medicine** was established in 1970 by the National Academy of Sciences to secure the services of eminent members of appropriate professions in the examination of policy matters pertaining to the health of the public. The Institute acts under the responsibility given to the National Academy of Sciences by its congressional charter to be an adviser to the federal government and, upon its own initiative, to identify issues of medical care, research, and education. Dr. Kenneth I. Shine is president of the Institute of Medicine.

The **National Research Council** was organized by the National Academy of Sciences in 1916 to associate the broad community of science and technology with the Academy's purposes of furthering knowledge and advising the federal government. Functioning in accordance with general policies determined by the Academy, the Council has become the principal operating agency of both the National Academy of Sciences and the National Academy of Engineering in providing services to the government, the public, and the scientific and engineering communities. The Council is administered jointly by both Academies and the Institute of Medicine. Dr. Bruce M. Alberts and Dr. Wm. A. Wulf are chairman and vice chairman, respectively, of the National Research Council.

COMMITTEE ON REVIEW AND EVALUATION OF THE ARMY NON-STOCKPILE CHEMICAL MATERIEL DISPOSAL PROGRAM

JOHN B. CARBERRY, *Chair*, E.I. duPont de Nemours and Company, Wilmington, Delaware
JOHN C. ALLEN, Battelle Memorial Institute, Washington, D.C.
RICHARD J. AYEN, Waste Management, Inc. (retired), Wakefield, Rhode Island
ROBERT A. BEAUDET, University of Southern California, Los Angeles
LISA M. BENDIXEN, Arthur D. Little, Inc., Cambridge, Massachusetts
JOAN B. BERKOWITZ, Farkas Berkowitz and Company, Washington, D.C.
JUDITH A. BRADBURY, Battelle Patuxent River, California, Maryland
MARTIN C. EDELSON, Ames Laboratory, Ames, Iowa
SIDNEY J. GREEN, TerraTek, Inc., Salt Lake City, Utah
PAUL F. KAVANAUGH, Consultant, Fairfax, Virginia
TODD A. KIMMELL, Argonne National Laboratory, Washington, D.C.
DOUGLAS M. MEDVILLE, MITRE Corporation (retired), Reston, Virginia
WINIFRED G. PALMER, Consultant, Frederick, Maryland
GEORGE W. PARSHALL, E.I. duPont de Nemours and Company (retired), Wilmington, Delaware
JAMES P. PASTORICK, Geophex UXO, Alexandria, Virginia
R. PETER STICKLES, Consultant, Concord, Massachusetts
WILLIAM J. WALSH, Pepper Hamilton LLP, Washington, D.C.
RONALD L. WOODFIN, Sandia National Laboratories (retired), Albuquerque, New Mexico

Board on Army Science and Technology Liaison

HENRY J. HATCH, U.S. Army (retired), Oakton, Virginia

Staff

NANCY T. SCHULTE, Senior Program Officer
DELPHINE D. GLAZE, Administrative Assistant
GREG EYRING, Consultant
DANIEL E.J. TALMAGE, JR., Research Associate

BOARD ON ARMY SCIENCE AND TECHNOLOGY

JOHN E. MILLER, *Chair*, Oracle Corporation, Reston, Virginia
GEORGE T. SINGLEY III, *Vice Chair*, Hicks and Associates, Inc., McLean, Virginia
ROBERT L. CATTOI, Rockwell International (retired), Dallas, Texas
RICHARD A. CONWAY, Union Carbide Corporation (retired), Charleston, West Virginia
GILBERT F. DECKER, Walt Disney Imagineering (retired), Glendale, California
ROBERT R. EVERETT, MITRE Corporation (retired), New Seabury, Massachusetts
PATRICK F. FLYNN, Cummins Engine Company, Inc. (retired), Columbus, Indiana
HENRY J. HATCH, U.S. Army (retired), Oakton, Virginia
EDWARD J. HAUG, University of Iowa, Iowa City
GERALD J. IAFRATE, North Carolina State University, Raleigh
MIRIAM E. JOHN, California Laboratory, Sandia National Laboratories, Livermore, California
DONALD R. KEITH, Cypress International (retired), Alexandria, Virginia
CLARENCE W. KITCHENS, IIT Research Institute, Alexandria, Virginia
SHIRLEY A. LIEBMAN, CECON Group (retired), Holtwood, Pennsylvania
KATHRYN V. LOGAN, Georgia Institute of Technology (professor emerita), Roswell, Georgia
STEPHEN C. LUBARD, S-L Technology, Woodland Hills, California
JOHN W. LYONS, U.S. Army Research Laboratory (retired), Ellicott City, Maryland
JOHN H. MOXLEY, Korn/Ferry International, Los Angeles, California
STEWART D. PERSONICK, Drexel University, Philadelphia, Pennsylvania
MILLARD F. ROSE, Radiance Technologies, Huntsville, Alabama
JOSEPH J. VERVIER, ENSCO, Inc., Melbourne, Florida

Staff

BRUCE A. BRAUN, Director
MICHAEL A. CLARKE, Associate Director
WILLIAM E. CAMPBELL, Administrative Coordinator
CHRIS JONES, Financial Associate
GWEN ROBY, Administrative Assistant
DEANNA P. SPARGER, Senior Project Assistant
DANIEL E.J. TALMAGE, JR., Research Associate

Preface

The Committee on Review and Evaluation of the Army Non-Stockpile Chemical Materiel Disposal Program (see Appendix A for biographies of committee members) was appointed by the National Research Council (NRC) to conduct studies on technical aspects of the U.S. Army Non-Stockpile Chemical Materiel Disposal Program. During its first year, the committee evaluated the Army's plans to dispose of chemical agent identification sets (CAIS)—test kits used for soldier training (NRC, 1999b). During the second year, the committee recommended nonincineration technologies that might be used for the posttreatment of neutralization wastes from Army non-stockpile materiel disposal systems (NRC, 2001a). During the third year, the Army asked the committee to supplement its report on neutralent wastes to include wastes produced by the Army's newest mobile system, the explosive destruction system (EDS) (NRC, 2001e). During this fourth year the committee has assessed the operational concepts for the mobile and semi-permanent facilities being developed by the product manager.

At its meetings, the committee was given a number of briefings (see Appendix B), and between meetings it held deliberations. The committee is grateful to the many individuals who provided technical information and insights during these briefings, particularly Lt. Col. Christopher Ross, Product Manager for Non-Stockpile Chemical Materiel, and his staff. This information provided a sound foundation for the committee's deliberations.

This study was conducted under the auspices of the NRC's Board on Army Science and Technology. The committee acknowledges the continued superb support of the director, Bruce A. Braun, as well as of NRC staff and committee members, who all worked diligently on a demanding schedule to produce this report.

John B. Carberry, *Chair*
Committee on Review and Evaluation
of the Army Non-Stockpile Chemical
Materiel Disposal Program

Acknowledgment of Reviewers

This report has been reviewed in draft form by individuals chosen for their diverse perspectives and technical expertise, in accordance with procedures approved by the NRC's Report Review Committee. The purpose of this independent review is to provide candid and critical comments that will assist the institution in making its published report as sound as possible and to ensure that the report meets institutional standards for objectivity, evidence, and responsiveness to the study charge. The review comments and draft manuscript remain confidential to protect the integrity of the deliberative process. We wish to thank the following individuals for their review of this report:

Elisabeth M. Drake, Massachusetts Institute of Technology (retired)
Gene Dyer, consultant
F. Wayne Jennings, consultant
Herbert J. Kouts, Defense Nuclear Facilities Safety Board (retired)
Richard Magee, Carmagan Engineering
James Michael, Environmental Protection Agency
Alvin Mushkatel, Arizona State University, and
William Tumas, Los Alamos National Laboratory

Although the reviewers listed above have provided many constructive comments and suggestions, they were not asked to endorse the conclusions or recommendations, nor did they see the final draft of the report before its release. The review of this report was overseen by John C. Bailar III, Professor Emeritus, University of Chicago. Appointed by the National Research Council, he was responsible for making certain that an independent examination of this report was carried out in accordance with institutional procedures and that all review comments were carefully considered. Responsibility for the final content of this report rests entirely with the authoring committee and the institution.

Contents

EXECUTIVE SUMMARY   1

1   BACKGROUND AND OVERVIEW   8
    The Stockpile Destruction Program, 9
        The Baseline Incineration Program, 9
        Alternative Technologies for Destroying the Stockpile, 9
        The Alternative Technologies and Approaches Program, 9
        The Alternative Technologies Program for Assembled Chemical
            Weapons Assessment, 9
    The Non-Stockpile Chemical Materiel Disposal Program, 10
        Non-Stockpile Sites, 10
        Non-Stockpile Inventory, 10
        Systems for Destroying NSCWM, 15
    Statement of Task, 16
    The Committee's Approach, 16
    Scope of the Report, 16
    Structure of the Report, 16

2   THE TOOLBOX OF NON-STOCKPILE TREATMENT OPTIONS   17
    Non-Stockpile Facilities, 17
        MAPS and PBNSF, 19
        Stockpile Facilities, 22
        Research and Development Facilities, 24
        Commercial Treatment, Storage, and Disposal Facilities,
    Mobile Treatment Systems, 27
        Explosive Destruction System. 28
        Rapid Response System, 29
        Single CAIS Accessing and Neutralization System, 31
        Donovan Blast Chamber, 32
    Individual Treatment Technologies, 34
        Plasma Arc, 34
        Chemical Oxidation, 37
        Wet Air Oxidation, 38
        Batch Supercritical Water Oxidation, 39
        Neutralization (Chemical Hydrolysis), 40
        Open Burning/Open Detonation, 42
        Integrated Ballistic Tent and Foam System, 43
        Multiple-Round Containers, 44

3 **APPLICATION OF THE NON-STOCKPILE TREATMENT SYSTEMS TO THE NSCWM INVENTORY**   46

   Introduction, 46
   Comparison of Candidate Technologies and Needs, 46
      CAIS PIGS, 48
      Individual CAIS Vials and Bottles, 48
      Small Quantities of Small Munitions, 49
      Chemical Agent in Bulk Containers, 49
      Binary Chemical Warfare Materiel Components, 49
      Unstable Explosive Munitions That Cannot Be Moved, 50
      Secondary Liquid Waste Streams, 50
      Large Quantities of NSCWM Items Currently in Storage, 51
      Large NSCWM Items, 51
      Large Quantities of Not-Yet-Recovered Small Munitions, 52
   NSCWM Treatment Categories for Which Available or In-Pipeline Tools Are Adequate, 52
      CAIS PIGs, 52
      Individual CAIS Vials and Bottles, 52
      Small Quantities of Small Munitions, 53
      Chemical Agents in Bulk Containers, 53
      Binary Chemical Warfare Materiel Components, 53
      Unstable Explosive Munitions That Cannot Be Moved, 53
      Secondary Liquid Waste Streams
   NSCWM Treatment Categories for Which Significant Additional Investment and Planning Are Needed, 54
      Large Quantities of NSCWM Items Currently in Storage, 54
      Large NSCWM Items, 54
      Large Quantities of Not-Yet-Recovered Small Munitions, 54
   Developing New Systems for New Finds, 54

4 **REGULATORY APPROVAL AND PERMITTING ISSUES**   56

   The Army's RAP Experience, 56
      Munitions Management Device, 56
      Rapid Response System, 57
      Spring Valley, Washington, D.C., 57
      Rocky Mountain Arsenal, Colorado, 57
      Munitions Assessment and Processing System, 57
      Pine Bluff Non-Stockpile Facility, 58
   Specific Issues, 58
      Regulatory Approval and Permitting Mechanisms, 58
      Cooperation between the Army, the States, and the Public in the RAP Process, 58
      Classification of Chemical Agent Identification Sets, 58
      Diverse Army Organizations with Responsibility for RAP, 58
      Schedule Requirements of the CWC, 59
      Overall Lack of a Regulatory Program for Treatment Requirements, 59
      Secondary Waste Classification, 60
      RAP for Mobile Technologies, 61
   Findings and Recommendations, 62
      Regulatory Permitting/Approval Mechanisms, 62
      Cooperation Between the Army, the States, and the Public in the RAP Process, 62
      Classification of CAIS, 62
      Diverse Army Organizations with a Responsibility for RAP, 62
      Schedule Requirements of the CWC, 63

Overall Lack of a Regulatory Program for Treatment Requirements, 63
Secondary Waste Classification, 63
RAP for Mobile Technologies, 63

5   PUBLIC INVOLVEMENT   64
Information Sources, 64
Stakeholder Views on Key Program Issues, 65
Stakeholder Views on Public Involvement, 66
NSCMP Planning for Public Involvement, 68
Findings and Recommendations, 69

REFERENCES   71

APPENDIXES
A   Biographical Sketches of Committee Members, 75
B   Committee Meetings and Other Activities, 79
C   Evaluation of the Suitability of Stockpile Chemical Disposal Facilities for Treating Stored Non-Stockpile CWM, 81
D   Non-Stockpile Facilities, 86
E   Mobile Non-Stockpile Systems, 91
F   Regulatory Background, 96
G   Transportation of Chemical Warfare Materiel, 103

INDEX   107

Tables and Figures

TABLES

1-1 Inventory of Non-Stockpile Items at Pine Bluff Arsenal, Pine Bluff, Arkansas, 11
1-2 Inventory of Non-Stockpile Items at Dugway Proving Ground (DPG) and Deseret Chemical Depot (DCD), Utah, 12
1-3 Inventory of Non-Stockpile Items at Aberdeen Proving Ground, Maryland, 13
1-4 Inventory of Non-Stockpile Items at Anniston Chemical Activity, Alabama, 13

2-1 Overview of Non-Stockpile Treatment Options, 18
2-2 Composition of Liquid Waste Streams from the EDS Treatment of Sarin (GB) Bomblets at RMA, 26
2-3 Numbers of Explosively Configured NSCWM and Total Recovered NSCWM, by Location, 29
2-4 NSCMP Technology Test Program, 35
2-5 Multiple-Round Containers and Their Contents, 44

3-1 Match of Primary Technologies and Systems to Items in the Non-Stockpile Inventory, 47
3-2 Focus of Secondary Technologies, 48

E-1 Approximate EDS Processing Time, by Agent, 94

FIGURES

1-1 Main chemical warfare agents in the U.S. inventory, 10

2-1 Floor plan of MAPS, 20
2-2 Glove-box system in the operations trailer of the RRS, 30
2-3 Schematic of one SCANS concept, 32
2-4 PLASMOX system layout, 36
2-5 Hydrolysis of the nerve agent GB (sarin), 41
2-6 Hydrolysis of DF with warm water, 42

D-1 Typical process flow for explosively configured munitions at PBNSF, 88
D-2 Typical process flow for non-explosively configured munitions at PBNSF, 89
D-3 Typical process flow for chemical agent identification sets at PBNSF, 89

E-1 Diagram of the EDS-1 vessel on its trailer, 93

Acronyms and Abbreviations

ABCDF	Aberdeen Chemical Agent Disposal Facility	DM	adamsite
ACW	assembled chemical weapons	DOD	U.S. Department of Defense
ACWA	Assembled Chemical Weapons Assessment (Program)	DOE	U.S. Department of Energy
		DOT	U.S. Department of Transportation
ANAD	Anniston Army Depot	DPG	Dugway Proving Ground
ANCDF	Anniston Chemical Disposal Facility	DRE	destruction and removal efficiency
APG	Aberdeen Proving Ground	DSHW	Division of Solid and Hazardous Waste (Utah)
ATAP	Alternative Technology Approach Program	DTV	drill-through valve
BGAD	Bluegrass Army Depot	ECC	explosive containment chamber
BGCDF	Bluegrass Chemical Disposal Facility	EDS	explosive destruction system
		EIS	environmental impact statement
CAC	Citizens' Advisory Commission	EPA	Environmental Protection Agency
CAIS	chemical agent identification set(s)	FOTW	federally owned treatment works
CAMDS	Chemical Agent Munitions Disposal System	FUDS	formerly used defense site
CDF	chemical disposal facility	GA	tabun (nerve agent)
CERCLA	Comprehensive Environmental Response, Compensation and Liability Act	GB	sarin (a nerve agent)
		GD	soman
CG	phosgene	H	sulfur mustard
CHATS	Chemical Agent Transfer System	$H\text{-}CHCl_3$	sulfur mustard in chloroform solution
CK	cyanogen chloride	HD	sulfur mustard (distilled)
CN	chloroacetophenone	HHS	Department of Health and Human Services
CSDP	Chemical Stockpile Disposal Program	HL	mustard-lewisite mixture
CTF	Chemical Transfer Facility	HN-1	nitrogen mustard 1
CWC	Chemical Weapons Convention	HN-3	nitrogen mustard 3
CWM	chemical warfare materiel	HS	sulfur mustard
CWWG	Chemical Weapons Working Group	HT	mustard agent T mixture
		HWIR	hazardous waste identification rule
D&D	drill and drain		
DBC	Donovan blast chamber	IMPA	isopropyl methylphosphonic acid
DCD	Deseret Chemical Depot	IRP	Installation Restoration Program
DCDMH	dichlorodimethylhydantoin		
DF	a binary precursor (methylphosphonic difluoride)	JACADS	Johnston Atoll Chemical Agent Disposal System

L	lewisite	PUCDF	Pueblo Chemical Disposal Facility
L-CHCl$_3$	lewisite in chloroform solution		
LDR	Land Disposal Restriction	QL	binary agent precursor (ethyl-2-diisopropylaminoethyl methylphosphonite)
MAPS	Munitions Assessment and Processing System		
MDM	multipurpose demilitarization machine	RAB	Restoration Advisory Board
MEA	monoethanolamine	RAP	regulatory approval and permitting
MPA	methylphosphonic acid	RCRA	Resource Conservation and Recovery Act
MPF	metal parts furnace		
MRC	multiple-round container	RCWM	recovered chemical warfare materiel
		R&D	research and development
NPL	National Priorities List	RDX	cyclotrimethylenetrinitramine
NRC	National Research Council	RMA	Rocky Mountain Arsenal
NS	non-stockpile	RRS	Rapid Response System
NSCM	Non-Stockpile Chemical Materiel		
NSCMP	Non-Stockpile Chemical Materiel Product	SBCCOM	Soldier and Biological Chemical Command
NSCWCC	Non-Stockpile Chemical Weapons Citizens' Coalition	SCANS	Single CAIS Accessing and Neutralization System
NSCWM	Non-Stockpile Chemical Warfare Materiel	SCWO	supercritical water oxidation
		SRC	single-round container
OB/OD	open burning/open detonation	TNT	trinitrotoluene
OPA	binary component (isopropyl alcohol with amine)	TOCDF	Tooele Chemical Agent Disposal Facility
		TSDF	treatment, storage, and disposal facility
OPCW	Organization for the Prohibition of Chemical Weapons	USACE	U.S. Army Corps of Engineers
		UCAR	Utah Chemical Agent Rule
PBA	Pine Bluff Arsenal	UV	ultraviolet
PBCDF	Pine Bluff Chemical Disposal Facility		
PBNSF	Pine Bluff Non-Stockpile Facility	VX	a nerve agent
PCD	Pueblo Chemical Depot		
PIG	package in-transit gas shipment	WAO	wet-air oxidation
PMCD	Program Manager for Chemical Demilitarization	WHEAT	water hydrolysis of explosives and agent technologies
PMNSCM	Product Manager for Non-Stockpile Chemical Materiel	WIPT	working integrated process team
POTW	publicly owned treatment works	3X	level of decontamination (suitable for transport or further processing)
PPM	parts per million		
PS	chloropicrin	5X	level of decontamination (suitable for commercial release)
PS-CHCl$_3$	chloropicrin in chloroform solution		

Executive Summary

The United States and other signatories of the Chemical Weapons Convention (CWC)[1] have committed to destroying all declared chemical warfare materiel (CWM) by April 29, 2007.[2] This materiel includes both stockpile materiel (all chemical agents and munitions available for use on the battlefield and stored at eight locations in the continental United States) and non-stockpile[3] materiel, a diverse category that encompasses all other CWM, which includes other chemical munitions and containers of chemical agent. Much of this non-stockpile chemical materiel (NSCM) was buried at current and former military installations in 31 states, the U.S. Virgin Islands, and the District of Columbia (U.S. Army, 1996). Only a small fraction of buried NSCM in the United States has been recovered. Virtually all NSCM that has been recovered is stored at stockpile storage sites.

The Army's baseline approach to destruction of stockpile CWM is to construct and operate state-of-the-art incinerators[4] at stockpile storage sites. However, incineration at any location, as well as transportation of agent of lethal intent across state lines, has met with strong public opposition and is subject to increasingly stringent regulatory requirements. The U.S. Army has developed or is investigating a mix of

---

[1] Formally known as the Convention on the Prohibition of the Development, Production, Stockpiling and Use of Chemical Weapons and on Their Destruction, the CWC requires the destruction of chemical weapons in the stockpile by 2007 and any non-stockpile weapons in storage at the time of the treaty ratification (1997) within 2, 5, or 10 years of the ratification date, depending on the type of chemical weapon or on the type of chemical with which an item is filled. Any chemical weapons "discovered . . . after the initial declaration of chemical weapons shall be reported, secured and destroyed in accordance with Part IV (A) of the Verification Annex" (CWC Article IV, Paragraph 9). Thus, non-stockpile CWM buried before January 1, 1997, is excluded from the treaty requirements as long as it remains buried. However, once this CWM is dug up and removed from the ground, the recovered CWM must be identified, declared under the CWC, inspected, and destroyed as soon as possible (U.S. Army, 2001a, pp. 1-3).

[2] Under the CWC, countries may apply for an extension of the deadline of up to 5 years. The United States has acknowledged that some of the stockpile destruction facilities are likely to continue to operate for several years beyond 2007. The Product Manager for Non-Stockpile Chemical Materiel (PMNSCM) has indicated to the committee that the PMNSCM intends to meet the 2007 deadline for destruction of all recovered non-stockpile materiel currently in storage.

[3] Non-stockpile chemical materiel includes buried chemical weapons, recovered chemical materiel, binary chemical weapons, former production facilities, and miscellaneous chemical materiel.

[4] The Johnston Atoll Chemical Agent Disposal System (JACADS), the initial stockpile facility, began destruction activities in 1990 and completed processing in November 2000 (U.S. Army, 2000a). There is an operating stockpile facility at Tooele, Utah, and facilities are undergoing systemization at Anniston, Alabama, and Umatilla, Oregon, and being constructed at Pine Bluff, Arkansas.

facilities and mobile systems that employ a variety of individual treatment technologies, including:

- the Rapid Response System (RRS), a mobile system for accessing and neutralizing the contents of chemical agent identification sets (CAIS)[5]
- the Single CAIS Accessing and Neutralization System (SCANS), a system for disposing of individual CAIS vials
- the Explosive Destruction System (EDS), a mobile system for destruction of all but the largest (by size, volume of agent, or energy of dispersing charge) chemical weapons
- the Donovan blast chamber (DBC), a system developed by a private company to treat conventional munitions, but which may have application to NSCM
- the single-round container (SRC) and the multiple-round container (MRC) for moving chemical materiel to a more suitable location when necessary
- two facilities, the Munitions Assessment and Processing System (MAPS) and the Pine Bluff Non-Stockpile Facility (PBNSF), based on cutting or drill-and-drain accessing of chemical agent, followed by chemical neutralization of agent and washing of hardware, arranged in a modular style and intended to process larger numbers of non-stockpile chemical weapons at a single site
- stockpile facilities, an additional option for the destruction of NSCWM[6] created in November 1999, when Congress amended the law to allow stockpile facilities to be used to destroy non-stockpile materiel
- a tent-and-foam system for partially contained detonation of a chemical weapon judged too sensitive to move
- various technologies that may be used to treat wastes resulting from the destruction of primary NSCM, known as secondary wastes

Before these systems can be operated, however, the Army must establish their technical effectiveness and safety, obtain the necessary regulatory approvals for operation at each site where NSCM is treated, and provide opportunities for public stakeholders to be involved in the decision-making process.

Accordingly, on March 16, 2001, the PMNSCM requested that the National Research Council (NRC) review the technical and operational plans for these facilities and mobile systems, make recommendations on their interrelationships, and assess the Army's plans for obtaining regulatory approvals and for enhancing public involvement in the decision-making process.

STATEMENT OF TASK

To help optimize the technical performance, as well as the regulatory approval and public acceptance processes, of the Non-Stockpile Chemical Materiel Disposal program, the NRC will:

- Evaluate mobile destruction systems and semi-permanent[7] facilities being used or considered by the Army's Non-stockpile product manager for the treatment of non-stockpile CWM and make recommendations on the systems and facilities that could be employed by the Army and their interrelationships. This analysis will specifically include consideration of issues and opportunities associated with the Explosive Destruction System (EDS), the Rapid Response System (RRS), the Munitions Assessment and Processing System (MAPS), the Pine Bluff Non-stockpile Facility (PBNSF), alternative treatments for neat chemicals, and selected aspects of the stockpile facilities.
- Review and evaluate the issues and obstacles associated with the environmental regulatory approval process for successful employment of Non-stockpile Chemical Materiel disposal systems (mobile and semi-permanent) that the Army may encounter during its management of the Non-stockpile Program and offer recommendations that may make the regulatory approval process more efficient while reducing schedule risk.
- Recommend areas in which further detailed study efforts would be particularly useful to the Product Manager.

Subsequently, it became clear to both the PMNSCM and the committee that the planning for the PBNSF and, to a lesser extent, the MAPS was not sufficiently evolved to allow an in-depth evaluation and analysis. At this writing, the PBNSF was still in the design phase, with key treatment technologies not yet selected; MAPS, while under construction, had not yet begun systematization and operational testing. Thus, the PMNSCM directed the committee to focus on the mobile systems and individual treatment technologies that will eventually be components of MAPS and PBNSF and to comment on how these components could be used efficiently in these two facilities.

APPROACH

This report begins by describing the non-stockpile chemical weapons materiel inventory, which contains unitary munitions and accessories dating back to World War I, binary munitions, and German munitions brought to the United States after World War II (Chapter 1). The non-stockpile inventory, which encompasses a greater variety of chemical agents than the stockpile inventory, includes blister, nerve, blood, and choking agents, as well as militarized industrial

---

[5]CAIS were used from 1928 to 1969 to train soldiers in the detection and identification of chemical agents.

[6]Until 2000, stockpile facilities were prohibited by law from accepting non-stockpile CWM; in P.L. 106-65, however, Congress amended the law to allow non-stockpile materiel to be destroyed in stockpile facilities, provided that the states in which the stockpile facilities are located agree.

[7]The U.S. Army Corps of Engineers defines semi-permanent facilities as having a life expectancy between 5 and 25 years. For the purposes of this report, the term "facilities" is used.

chemicals and binary agents, and its condition is highly variable (some items have severely deteriorated during decades of burial).

The committee then assesses the tools, or specific options, available to PMNSCM to safely destroy these items. These tools, which include facilities, mobile treatment systems, and individual treatment technologies, are evaluated in Chapter 2 from the standpoint of their current status, as well as technical, regulatory, and permitting (RAP) and public involvement issues. The Army has prepared assessments of the potential health and environmental impacts of its transportable treatment systems (U.S. Army, 2001a), including risks during normal operations and from accidental release of hazardous substances. Similar site-specific assessments are generally required of the Army's treatment facilities as part of the permitting process. The committee did not review the specific methodology or the regulatory assumptions used by the Army in assessing these health and environmental impacts because the overall risk assessment methodologies are of the kind typically used in U.S. regulatory and permitting programs.

In Chapter 3, the committee matches the treatment options with the materiel or munitions to be treated, identifies gaps in the program, and makes recommendations on the facilities, systems, and technologies. Chapter 4 examines RAP issues for waste management and identifies issues that, when resolved, will facilitate the RAP process in the future. Chapter 5 commends PMNSCM for its increased openness in providing information to a range of stakeholders and in developing relationships with them and notes areas that might be improved. Throughout the report, findings and recommendations follow each discussion. Appendix C evaluates the suitability of stockpile chemical disposal facilities for treating stored non-stockpile facilities. More detailed information about MAPS and PBNSF, two non-stockpile facilities, one under construction (MAPS) and one in design (PBNSF), is provided in Appendix D. Appendix E reviews the RRS and the EDS, two mobile systems for treatment of NSCWM. Appendixes F and G provide background information on regulatory and permitting issues and transportation of CWM, respectively.

Because the Army appears to be making excellent progress in destroying old production facilities, empty ton containers, and unfilled CWM delivery systems, these categories are not discussed in this report. Instead, the report covers subcategories of NSCWM whose destruction appears to pose the greatest challenges, including CAIS, recovered chemical munitions, binary CWM components, and chemical agent in bulk containers. Treatment of secondary waste streams generated in the treatment of this Non-Stockpile Chemical Warfare Materiel (NSCWM) is considered along with the treatment of primary waste.

The committee concurs with reports issued by other NRC committees (e.g., NRC, 1994) and reaffirms its own previous reports (e.g., NRC, 1999a)—namely that state-of-the-art incineration is safe and effective for the destruction of chemical weapons agent and energetics. However, the committee also recognizes that widespread opposition to incineration has led to considerable delays and additional costs. For that reason, it has worked with the Army to help evaluate alternatives to direct incineration.

The committee considered 10 categories of NSCWM that the Army currently faces or is likely to face in the future and examined the adequacy of the available treatment tools:

1. CAIS packages for in-transit gas shipment (PIGs)[8]
2. individual CAIS vials and bottles
3. small quantities of small munitions
4. chemical agent in bulk containers
5. binary chemical warfare materiel components
6. unstable explosive munitions that cannot be moved
7. secondary liquid waste streams
8. large quantities of NSCWM items currently in storage
9. large NSCWM items
10. large quantities of not-yet-recovered small munitions

The committee found that for the first seven categories, the Army has tools available or under development that should enable the destruction of NSCWM in an effective and timely way. However, significant additional investment or planning will be required to satisfactorily address the issues posed by the final three categories. Key recommendations relating to these categories appear below. The underlying discussions and findings, as well as additional findings and recommendations, appear in Chapters 1 through 5.

KEY RECOMMENDATIONS

The following findings and recommendations do not in any way diminish the committee's previous findings that state-of-the-art incineration is safe, robust, and effective (NRC, 2001a). PMNSCM has already invested considerable resources in developing treatment options to address many of the NSCWM treatment contingencies it may face. In some cases, this investment has yielded treatment systems that are ready for use; in others, treatment systems that are currently in the development pipeline should, upon completion, offer adequate capabilities.

The non-stockpile program also has available to it treatment facilities that were developed for the stockpile program, as well as commercial hazardous waste disposal facilities. To adapt these facilities for the treatment of NSCWM secondary wastes, equipment modifications or permit modifications may be required, but the technical feasibility seems

---

[8]PIGs are metal canisters with packing material designed to protect CAIS during transport.

clear. The committee's findings and recommendations relating to the 10 NSCWM treatment categories are discussed further below.

Ten Treatment Categories

CAIS PIGs

Finding: The RRS is an expensive but adequate treatment system for CAIS PIGs and large numbers of loose CAIS vials and bottles. As other treatment options are also possible, this category appears to be well covered (Finding 2-7).

Individual CAIS Vials and Bottles

PMNSCM is developing the single CAIS accessing and neutralization system (SCANS) to treat individual CAIS vials and bottles recovered at remote sites. When fully developed, this system should be well suited to this task.

Recommendation: The committee recommends that PMNSCM continue to develop and optimize SCANS to increase the number of CAIS vials and bottles that can be cost-effectively treated with multiple SCANS units. If the development program results in a system that can be cost-effectively used for a large number of vials and bottles, the system should be fielded as rapidly as possible. This approach would allow reserving the RRS for treating very large numbers of CAIS and PIGs containing CAIS, which the SCANS cannot process (Recommendation 2-8).

Small Quantities of Small Munitions

PMNSCM has developed the transportable EDS[9] as the workhorse system for destruction of both explosively and nonexplosively configured munitions in the field. The EDS-1 prototype was recently deployed to Rocky Mountain Arsenal, where it successfully destroyed 10 sarin bomblets. Improved versions of the EDS-1 as well as a larger EDS-2 are currently in development. Once these developments have been completed, it appears that this category will be well covered. The EDS system appears to be sufficiently flexible that it might also be used for other NSCWM treatment categories.

Recommendation: The committee recommends that the Army continue to implement the planned improvements of the EDS that increase explosive capacity and reduce pro-

---

[9]The EDS was originally developed to destroy non-stockpile items that were deemed too unstable for transport or long-term storage; however, it can also be used to treat limited numbers of stable chemical munitions, with or without explosive components.

cessing cycle time. The Army should consider the applicability of the EDS as modules in facilities (Recommendation 2-5).

Chemical Agent in Bulk Containers

The non-stockpile inventory includes numerous containers of chemical agents of various types and sizes that have accumulated over the years. In general, these are stored at stockpile sites. There are many treatment options available for these bulk containers; the most obvious is to use the stockpile chemical disposal facilities (CDFs), although modifications may be required and permit modifications may be difficult to obtain.

In addition to the stockpile facilities, two experimental facilities have long been used to destroy a variety of chemical agents by chemical neutralization: these are the Chemical Transfer Facility (CTF) at Aberdeen Proving Ground, Maryland, and the Chemical Agent and Munitions Destruction System (CAMDS), at Deseret Chemical Depot, Utah. Although these are R&D facilities and therefore should not be used on a routine basis to destroy NSCWM, they might be considered as an option to destroy limited numbers of non-stockpile items that contain unusual chemical fills or that have a configuration that cannot be handled by other systems.

Further treatment options for non-stockpile bulk chemicals include direct destruction in a plasma arc system (see below) or even treatment in the EDS. With all of these options available, this category is well covered.

Recommendation: While recognizing that there are significant regulatory and public acceptability issues to resolve, the committee recommends that non-stockpile chemical materiel in bulk containers located at stockpile sites and suitable for destruction in chemical stockpile disposal facilities be destroyed in those facilities (Recommendation 3-1).

Binary Chemical Warfare Materiel Components

The entire non-stockpile inventory of binary CWM components is stored in canisters and drums at Pine Bluff Arsenal, a stockpile site. Options for treatment include destruction in the Pine Bluff Chemical Disposal Facility, direct destruction in a plasma arc system, or chemical neutralization followed by oxidative posttreatment of the neutralents. The high concentration of fluorine in the binary component DF raises concerns about corrosion in some treatment systems.

Recommendation: Additional testing of plasma arc technology should be done to ensure that proposed plasma arc systems are capable of meeting requirements of the Environmental Protection Agency (EPA) and state requirements (Recommendation 2-10).

Recommendation: Ideally, the binary precursors methylphosphonic difluoride (DF) and ethyl-2-diisopropylaminoethyl methylphosphonite (QL) stored at Pine Bluff Arsenal should be destroyed directly, either by burning in the Pine Bluff Chemical Destruction Facility incinerator or by plasma arc treatment. If these facilities cannot handle the fluorine-rich DF destruction products, the committee recommends that on-site neutralization followed by oxidative post-treatment of the neutralents be developed. The easiest post-treatment may be shipment to a commercial incinerator capable of dealing with high levels of fluorine (Recommendation 3-2).

Unstable Explosive Munitions That Cannot Be Moved

Open burning/open detonation (OB/OD) has been the traditional method of disposing of unstable munitions, including chemical munitions, but OB/OD is no longer considered acceptable for NSCWM by regulators except in emergencies. The Army has been exploring an alternative to OB/OD called the tent-and-foam system, which provides for partially contained detonation of unstable munitions.

Recommendation: The Army should complete the development and testing of the tent-and-foam system for controlling on-site detonation of unstable munitions (Recommendation 3-3).

Secondary Liquid Waste Streams

Treatment systems such as the RRS and EDS that rely on chemical neutralization of agents produce secondary liquid waste streams of two types:

- neutralent waste streams consisting largely of organic solvents and agent neutralization by-products
- aqueous waste streams, including rinsates, washes, and brine solutions

The Army's plan for destruction of these wastes involves the collection of neutralized agent (neutralent), washes, and rinsates followed by treatment on-site or shipment to a commercial or federal treatment, storage, and disposal facility (TSDF) for final disposal. Disposal of these neutralents, washes, and rinsates would generally be by incineration. However, at least some of these liquids may be suitable for destruction by other technologies, existing or yet to be demonstrated. PMNSCM has undertaken a technology test program to test a large number of alternative technologies for destruction of these secondary waste streams.

Recommendation: The PMNSCM should continue its research and development program on chemical oxidation and wet air oxidation of neutralents and rinsates (Recommendation 2-12a).

Recommendation: Consistent with the committee's earlier analyses (NRC, 2001a, 2001b), there should be no further funding for the development of biological treatments, electrochemical oxidation, gas-phase chemical reduction, solvated electron technology, and continuous SCWO technologies for the treatment of neutralents and rinsates. PMNSCM should monitor progress in technologies being developed under the Assembled Chemical Weapons Assessment (ACWA) program but should evaluate ACWA technologies for the treatment of non-stockpile neutralents and rinsates only if no additional investment is required (Recommendation 2-12b).

In the following areas, the committee judges the treatment options that are available or in the pipeline to be insufficient to permit the non-stockpile program to meet its goals. Additional investment or planning efforts are needed.

Large Quantities of NSCWM Items Currently in Storage

Some 85 percent of all recovered NSCWM in the United States is stored at Pine Bluff Arsenal. The Army has under design the Pine Bluff Non-Stockpile Facility (PBNSF) to destroy this material, with the assistance of an RRS and an EDS to treat CAIS and certain explosively configured munitions, respectively.

Recommendation: PMNSCM should develop a detailed, realistic timetable showing how the planned non-stockpile facilities at Pine Bluff Arsenal can achieve the throughput necessary to destroy the stored non-stockpile items by April 2007 and should communicate this timetable to all stakeholders (Recommendation 2-1).

Large NSCWM Items

Disposal of chemical projectiles larger than 155 mm and large (500 or 1,000 lb) bombs presents a special challenge to the non-stockpile program. Although such munitions are rarely recovered in the United States, they have been recovered as a result of U.S. activities in at least one foreign country, and it is likely they will be found on U.S. soil in the future.

Recommendation: PMNSCM should develop a strategy for treating chemical bombs and projectiles that are too large for treatment in the EDS, in the DBC (if successfully demonstrated), or in planned facilities. One option is to test the British drill-through valve (DTV) system, modify it if necessary, and prepare it for use on existing large NSCWM items and other such items that may be found in the future (Recommendation 3-4).

Large Quantities of Not-Yet-Recovered Small Munitions

Sites at which thousands of NSCWM items are believed to be buried present a special challenge to the non-stockpile program. Examples of such sites include Deseret Chemical Depot, Utah; Rocky Mountain Arsenal, Colorado; and Redstone Arsenal, Alabama. Use of one or even a few EDS units would be inefficient given their relatively low throughput capacity (currently one munition every 2 days). At present, the Army's only option for cleaning up such a site would be the construction of a facility such as MAPS or PBNSF. However, such facilities are expensive and have a large environmental footprint. A transportable treatment system with a high throughput would be highly desirable to treat this category of NSCWM.

Recommendation: The non-stockpile program should continue to monitor the Belgian tests of the DBC. If the results are encouraging and it appears that the DBC can be permitted in the United States, it should be considered for use at sites where prompt disposal of large numbers of munitions is required (Recommendation 2-9).

Regulatory Approval and Permitting

Historically, establishing regulatory approval and permitting (RAP) requirements for new systems and technologies has been shown to be a resource-intensive and time-consuming process. Obtaining regulatory approvals is likely to be a critical factor in meeting the treaty deadlines for the destruction of NSCWM. Communication and cooperation with regulators (particularly state regulators), combined with an effective public involvement program, are essential for obtaining regulatory approvals in a timely manner. The committee urges the Army and states to enhance the existing cooperative efforts to define appropriate regulatory requirements for the technologies.

Communicate with Regulators

Recommendation: The Army should establish a pre-permitting process to resolve RAP issues involving the Army, regulators, and the public for both mobile systems and non-stockpile treatment facilities. In addition, the Army should develop guidance on RAP for management of NSCWM. A guidance that is jointly issued by the Army and regulators, with input from the public, should be considered, and the committee recommends that it be of national scope (Recommendation 4-2).

Recommendation: The Army should examine funding provided to the states as part of existing cooperative agreements to ensure that they are sufficient to evaluate new or innovative NSCWM treatment technologies within a time frame consistent with CWC deadlines (Recommendation 4-5).

Recommendation: The Army and the states should continue to work together to achieve mutually acceptable regulations that define appropriate treatment for chemical agents and associated wastes. While state-specific treatment standards can be established, the committee recommends standards that are national in scope (Recommendation 4-6).

Develop Solid Working Relations

Recommendation: The Army should work with state regulators to tailor RAP mechanisms to the magnitude of the NSCWM recovery and treatment operations. For facilities, initial operations should be conducted under expedited RAP mechanisms (e.g., a Research, Development, and Demonstration permit); traditional Resource Conservation and Recovery Act (RCRA) permits, if necessary, should be employed after operations become routine. When mobile treatment systems or technologies are employed, and particularly for small or even moderate quantities of newly discovered NSCWM, expedited (non-RCRA permit) regulatory approval mechanisms under RCRA or the Comprehensive Environmental Response, Compensation and Liability Act (CERCLA) should be used, as appropriate (Recommendation 4-1).

Recommendation: RAP for all of the Army's chemical agent programs, including the non-stockpile program, should be seamless and transparent to the regulator and the public, who should "see" only one Army across all chemical agent programs at a specific location or operation. An installation-specific (or in the case of off-site NSCWM finds, operation-specific) core Army RAP team should be established for all chemical agent operations, including treatment of NSCWM. Installation or operation representatives should lead the RAP team at each location. The team should be directed by a central Army organization encompassing all chemical agent operations that require RAP so as to promote communication, continuity, and consistency among them. This organization should have the authority to establish RAP policy for all chemical agent operations nationwide (Recommendation 4-4).

Avoid Unnecessary Conservatism

Recommendation: The Army should reverse its classification of CAIS as recovered chemical warfare materiel (RCWM), thus avoiding additional time and cost for their destruction (Recommendation 4-3).

Recommendation: In states where secondary waste streams are regulated as acutely hazardous, the Army should work with state regulators to remove the designation "acutely hazardous." For neutralents, the Army should work with state regulators to establish de minimis concentrations for the agents in waste streams, to be incorporated into the listing regulations, whereby the waste would no longer be considered as being associated with the parent agent waste. Further, the Army and the states should consider whether rinsates and cleaning solutions and residuals from the treatment of neutralent should be classified as hazardous waste at all (Recommendation 4-7).

Recommendation: Given the similarities between NSCWM secondary wastes and industrial hazardous wastes, the committee recommends that no additional prohibitions be placed on the off-site transportation of secondary wastes (Recommendation 4-8).

Public Involvement

As noted in the committee's three previous reports (NRC, 1999a, 2001a, 2001b), it is necessary and desirable that the Army proactively seek public involvement in policy decisions that once were considered to require only scientific judgment.

Recommendation: As with RAP activities, public involvement should appear seamless across Army programs and transparent to local and national stakeholders. The committee recommends that the Army establish central direction to ensure coordination of program and installation missions and to promote continuity and consistency in public involvement programs across installations and between program and installation staff (Recommendation 5-1).

Recommendation: The committee recommends that the Army expand its public affairs program to include involvement as well as outreach activities.[10] Specifically, for the Army to gain from lessons documented in studies of the stockpile program, the committee recommends as follows:

- The Army should direct installations to implement, in coordination with program staff, a strategy that includes development of public involvement mechanisms. Such mechanisms must be fully integrated with project schedules so that the public has a genuine opportunity to provide input to project decisions. Their goal must be to engage both the local public and other stakeholders in discussing and evaluating the various technologies being considered and to provide a continuing means of involving them in future planning efforts and project decisions.
- The Army should conduct public involvement training for program and installation personnel, including commanders, public relations, and program technical staff. Such training must be more extensive than a one-day training course in risk communication and must be conducted very early in the program. The training should be provided on a continuing basis to ensure adequate preparation of newly assigned personnel.
- NSCMP should consider how the program could more effectively use existing mechanisms, such as the Core Group, to include and engage citizens at the local, site-specific level as well as at the national level in identifying specific concerns and actively contributing to consideration of the trade-offs inherent in program decisions (Recommendation 5-2).

---

[10]These components are generally consistent with the threefold division of public affairs provided in a letter report from the Committee on Review and Evaluation of the Army Chemical Stockpile Disposal Program. The components were public relations (provision of written information materials), outreach (opening channels of communication to the public so that their values, concerns, and needs can be heard), and involvement (development of a formal process that gives stakeholders an opportunity to provide input to decisions without surrendering the agency's legal mandate to make those decisions) (NRC, 2000a).

1

Background and Overview

Since World War I, the United States has considered it necessary to have the capability to engage in chemical warfare. Today, as a result of the United States' decision to sign and ratify the Chemical Weapons Convention (CWC),[1] the long-term storage of aging chemical warfare materiel (CWM) is no longer permitted. Also, the public is concerned about the risks associated with the long-term storage of CWM. Consequently, the United States and other signatories of the CWC are in the process of destroying all declared[2] CWM by the treaty deadline of April 29, 2007.[3]

U.S. law and international treaties have divided CWM into two categories: "stockpile" and "non-stockpile." Stockpile materiel includes all chemical agents available for use on the battlefield, including chemical agents assembled into weapons and in bulk (1-ton) containers. Stockpile materiel is stored at eight locations in the United States.

Non-stockpile materiel is a diverse category that includes all *other* chemical weapon-related items.[4] Much of this materiel was buried on current and former military sites but is now being recovered as the land is remediated. Some CWM also is buried at current and former test and firing ranges. Non-stockpile materiel that has been recovered is now stored at several military installations across the United States. According to the CWC, non-stockpile CWM items in storage at the time of treaty ratification in April 1997 must be destroyed within 2, 5, or 10 years, depending on the type of chemical weapon and the type of agent. Non-stockpile CWM recovered after treaty ratification must be declared under the CWC and destroyed "as soon as possible" (U.S. Army, 2001b). Generally, non-stockpile items that are recovered have been transported to a nearby stockpile site for safe storage.[5]

The U.S. Army's Program Manager for Chemical Demilitarization (PMCD) has overall responsibility for dispos-

---

[1] Formally, the Convention on the Prohibition of the Development, Production, Stockpiling and Use of Chemical Weapons and Their Destruction. The treaty was signed by the United States on January 13, 1993, and ratified by the U.S. Congress on April 25, 1997. The CWC specifies the time period within which covered categories of CWM must be destroyed.

[2] CWM that remains buried is not subject to the treaty. Once the CWM has been recovered and characterized, it must be declared under the CWC and must then be destroyed as soon as possible. Non-stockpile CWM items in storage at the time of treaty ratification must be destroyed within 2, 5, or 10 years of the ratification date, depending on the type of chemical weapon or on the type of chemical with which an item is filled (U.S. Army, 2001c).

[3] Under the CWC, countries may apply for an extension of the deadline of up to 5 years. The United States has acknowledged that some of the stockpile destruction facilities are likely to continue to operate for several years beyond 2007. The Product Manager for Non-Stockpile Chemical Materiel has indicated to the committee that the NSCMP intends to meet the 2007 deadline for destruction of all recovered non-stockpile materiel currently in storage.

[4] The category includes buried chemical warfare materiel, recovered chemical warfare materiel, binary chemical weapons, former production facilities, and miscellaneous chemical warfare materiel.

[5] An exception is recovered chemical agent identification sets (CAIS), which contain small quantities of chemical agents and militarized industrial chemicals, used for training purposes. These are sometimes stored at the site where they are recovered.

ing of all CWM. Under the PMCD's direction are two programs: the Chemical Stockpile Disposal Program (CSDP) and the Non-Stockpile Chemical Materiel Product (NSCMP).[6]

Although this study is concerned with the destruction of non-stockpile materiel, a brief review of the stockpile destruction program is presented here, because (1) the latter program is more mature than the non-stockpile destruction program and (2) the chemical agents in the two inventories are similar, so that many of the technologies and social or political factors that have influenced the stockpile program are expected to have similar effects on the non-stockpile program.

THE STOCKPILE DESTRUCTION PROGRAM

The Baseline Incineration Program

The U.S. effort to destroy its stockpile chemical materiel was already well under way at the time the CWC was signed, in January 1993.[7] The Army's baseline method for destroying the stockpile is to construct chemical disposal facilities in which the chemical agents are incinerated. There is currently one operating chemical disposal facility in the continental United States, at the Deseret Chemical Depot (DCD) near Tooele, Utah. An additional facility on Johnston Atoll in the Pacific Ocean completed destruction of its inventory in 2000 and is presently undergoing closure.[8] Together, these two facilities are expected to destroy about one-half of the U.S. stockpile, the remainder of which is dispersed among seven other storage sites in the continental United States.

Because federal law prohibits the shipment of these weapons across state lines for disposal,[9] the Army planned to construct similar incineration systems at other sites. Baseline (incineration) facilities have been completed, and operational testing is under way in Umatilla, Oregon, and Anniston, Alabama. In Pine Bluff, Arkansas, construction of a chemical disposal facility was approximately two-thirds complete as of March 2002.[10]

Until 2000, stockpile facilities were prohibited by law from accepting and destroying non-stockpile CWM; in November 1999, however, Congress amended the law (P.L. 106-65) to allow non-stockpile materiel to be destroyed at stockpile facilities, provided that regulatory authorities in the states where the stockpile facilities are located agree.

Alternative Technologies for Destroying the Stockpile

The Army's choice of incineration as a disposal technology has met with strong public and political opposition at some locations—so much so that the Army has sought alternative nonincineration technologies for destroying stockpile chemical agents in two key programs: one for chemical agents stored in 1-ton bulk containers and one for chemical agents in assembled chemical weapons.

The Alternative Technologies and Approaches Program

In the Alternative Technologies and Approaches Program (ATAP), neutralization processes based on hydrolysis of chemical agent either in pure water or in sodium hydroxide solution have been developed to destroy the chemical agents in bulk (1-ton) containers stored at Aberdeen, Maryland, and Newport, Indiana. Construction of facilities to carry this out is under way (NRC, 1994, 1998).

The Alternative Technologies Program for Assembled Chemical Weapons Assessment

In 1996, Congress also mandated and appropriated money for the Army to demonstrate at least two nonincineration technologies for the destruction of assembled chemical weapons (ACW) for possible use at the Pueblo Chemical Depot (PCD), in Colorado, and at the Bluegrass Army Depot (BGAD), in Kentucky. Six technologies passed the initial screening, and three were finally selected for demonstration (NRC, 2000b). Two, the Parsons/Honeywell water hydrolysis of explosives and agent technologies (WHEAT) process and General Atomics Total Solution (GATS), were further evaluated in an engineering design study at Pueblo Chemical

---

[6]As discussed in Chapter 4, many different Army organizations are involved in making decisions about sites where non-stockpile chemical materiel is stored or recovered. Often, the site is a current military base or depot, where the primary decision maker is the base or depot commander. The role of the NSCMP is to develop treatment technology options and offer treatment services to decision makers at these sites.

[7]In November 1985, Congress passed Public Law 99-145, requiring destruction of stockpile agents and munitions.

[8]Chemical weapons stored overseas were collected at Johnston Island, southwest of Hawaii, and destroyed by the Johnston Atoll Chemical Agent Disposal System (JACADS), the initial chemical demilitarization facility. JACADS began destruction activities in 1990 and completed processing of more than 2,000 tons of agent and more than 400,000 munitions and containers in the overseas stockpile in November 2000 (U.S. Army, 2000a).

[9]P.L. 103-337 prohibited the transport across state lines of CWM in the stockpile; it allowed *regulated* movement of non-stockpile chemical materiel across state lines.

[10]The stockpile inventories at Aberdeen Proving Ground, Maryland, and at Newport Chemical Depot, Indiana, consist of bulk agents that will be destroyed by chemical hydrolysis. Technologies for destruction of the stockpile inventories at Bluegrass Army Depot, Kentucky, and Pueblo Chemical Depot, Colorado, have not yet been selected.

Depot. Congress then mandated that the program manager for ACW demonstrate the remaining three undemonstrated technologies. These last three technologies were demonstrated from March to July 2000 (NRC, 1999b, 2000b, 2001c). Two of them, AEA's Silver II process and the Eco-Logic/Foster Wheeler process, were selected to be carried forward to an engineering design study. These two technologies, along with GATS, are candidates for destroying the chemical munitions at Bluegrass Army Depot.

THE NON-STOCKPILE CHEMICAL MATERIEL DISPOSAL PROGRAM

Prior to 1991, the U.S. effort to dispose of CWM was limited to stockpile materiel. The 1991 Defense Appropriations Act directed the Secretary of Defense to establish the Product Manager for Non-Stockpile Chemical Materiel (PMNSCM) with responsibility for the destruction of non-stockpile CWM.

Non-Stockpile Sites

In the 1993 Defense Appropriations Act,[11] Congress directed the Army to report the locations, types, and quantities of non-stockpile chemical materiel; to report the methods to be used for its destruction; to provide cost and time estimates; and to assess transportation options. A survey and analysis report provided an overview of the task facing the Army (U.S. Army, 1993, updated in draft form in 1996).

As of 1996, the Army had located 168 potential burial sites at 63 locations in 31 states, the U.S. Virgin Islands, and the District of Columbia (U.S. Army, 1996). Of the 63 locations, most are current or former military facilities. They include (1) sites with chemical agent identification sets (CAIS) only, (2) sites with small quantities of materiel with no associated explosives, (3) sites with small quantities of materiel with explosives, and (4) sites with large quantities of materiel with and without explosives. The majority of the sites involve small quantities of materiel.

Non-Stockpile Inventory

Non-stockpile chemical weapons materiel (NSCWM) is far more diverse than stockpile CWM: for example, it contains U.S. unitary munitions and accessories dating back to World War I, binary munitions, and German munitions brought back to the United States after World War II. There is a greater variety of chemical agents in NSCWM than in stockpile materiel (including blister agents, nerve agents, blood agents, and choking agents),[12] as well as militarized industrial chemicals. Energetics found in chemical munitions include aromatic nitro compounds such as trinitrotoluene (TNT), aromatic nitramines such as tetryl, heterocyclic nitramines such as cyclotrimethylenetrinitramine (RDX) and high-melting explosive (HMX), and nitrate esters used in propellants (e.g., nitrocellulose and nitroglycerine). The most commonly encountered energetics are tetryl, TNT, and composition B (60 percent RDX, 39 percent TNT, 1 percent beeswax).[13] The condition of the NSCWM is also much more variable than that of the stockpile, especially for items that have severely deteriorated after being buried for decades.

The Chemical Weapons Convention (CWC) requires that the following non-stockpile categories already declared at the time of the CWC's entry into force in 1997 be destroyed by April 29, 2007: binary CWM components; recovered CWM (excluding CAIS, which were developed for defensive purposes and were not intended to be lethal, and are therefore not covered by the CWC); former production facilities; and some types of miscellaneous CWM, including containers filled with agent that was removed from leaking munitions. The major chemical agents in the U.S. inventory are phosgene (CG, $COCl_2$) and the compounds shown in Figure 1-1: GB (sarin), VX, and HD (mustard). The chemicals DF (CH_3POF_2) and QL ($CH_3P(OEt)(OCH_2CH_2N\text{-}i\text{-}Pr_2)$), which are precursors for GB and VX, respectively, are also components.

Tables 1-1 through 1-4 present the most current information available to the committee regarding the numbers, types of agent fills, and explosive configurations of recovered mu-

---

[11]Section 176 of P.L. 102-484.

[12]John Gieseking, Office of the PMNSCM, presentation to the committee on June 16, 1999.

[13]Stone & Webster information paper briefed to the committee on October 14, 1999.

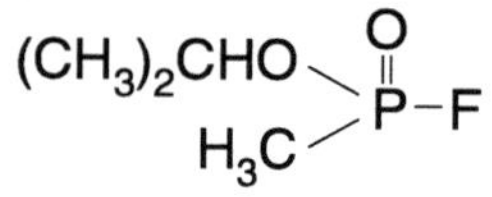

$(CH_3)_2CHO\diagdown\underset{H_3C\diagup}{\overset{\overset{O}{\|}}{P}}-F$

GB (sarin)

$\underset{CH_3CH_2O\diagup}{\overset{CH_3\diagdown}{\overset{\overset{O}{\|}}{P}}}-S-CH_2CH_2-N\diagdown\overset{CH(CH_3)_2}{\diagup CH(CH_3)_2}$

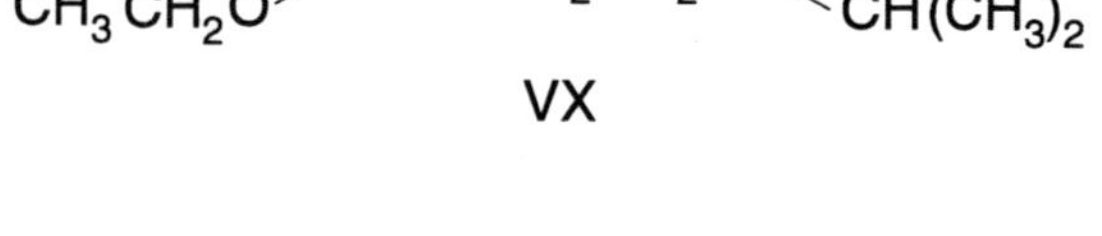

VX

$(ClCH_2CH_2)_2S$

HD (mustard)

FIGURE 1-1 Main chemical warfare agents in the U.S. inventory. SOURCE: NRC (1994).

TABLE 1-1 Inventory of Non-Stockpile Items at Pine Bluff Arsenal, Pine Bluff, Arkansas

Item	Chemical Fill								Total No. of Items
	H/MD/HN/ HS/HT	GA/GB/GD	VX	DM/L	CG/CK	DF	QL	Other	
Munition									
Explosive									
4.2-inch mortar round	727			2	1				730
75-mm projectile	16								16
200-mm Livens projectile	12				3				15
155-mm projectile	1								1
105-mm projectile	1								1
M70A1 bomb (poss. explosive)	9								9
150-mm Traktor rocket/warhead (HN-3)	479								479
Nonexplosive									
75-mm projectile	3								3
Subtotal	1,248			2	4				1,254
Chemical sample container[a]									
Ton container		2							2
4-inch cylinder	1								1
Lab sample container			2						2
Vial (L)				1					1
Subtotal	2	2	2						6
Chemical agent ID set (CAIS)									
Mustard (H/HD/HS)	5,764								5,764
Nitrogen mustard (HN-1 and -3)	50								50
Lewisite (L)				397					397
Chloropicrin (PS)								396	396
Phosgene (CG)					396				396
Chloroacetophenone (CN)								17	17
Adamsite (DM)				17					17
Triphosgene (TP)								17	17
Cyanogen chloride (CK)					33				33
Diethyl malonate, etc. (GS)								33	33
Subtotal	5,814			414	429			463	7,120
Binary agent precursor									
M20						56,820			56,820
Drum						7	293		300
Box, container, can							3		3
Subtotal						56,827	296		57,123
Empty ton container[b]				4,375					4,375
Total	7,063	2	2	4,789	433	56,827	296	463	69,878

[a]Inventory consists of individual CAIS items, not complete CAIS.
[b]Sampling of some of these containers indicated that they may be contaminated with lewisite, arsenic, and/or mercury.

SOURCE: Provided to the committee by PMNSCM on July 10, 2001.

TABLE 1-2 Inventory of Non-Stockpile Items at Dugway Proving Ground (DPG) and Deseret Chemical Depot (DCD), Utah[a]

Item	Location	Chemical Fill H/HD/HN/HT/HS	GA/GB/GD	Lewisite	VX/Vx	Total No. of Items
Munitions						
Explosive						
4.2-inch mortar round	DPG	10		12[b]		22
105-mm projectile	DPG	2		4[b]		6
155-mm projectile	DPG		1			1
T77 155-mm projectile	DPG		1			1
6-inch projectile	DPG		1			1
M-125 bomblet	DPG		1			1
Nonexplosive						
155 mm (1 empty but contaminated)	DPG		2			2
4.2-inch mortar round	DPG	2				2
M-139 half bomblet	DPG		1			1
Subtotal		14	7	16		37
Chemical sample containers						
Ton container	DCD	1				1
Container, bottle, vial	DPG	8[c]	48[d]	1	28	85
Bottle (VX (EA-1699))	DPG				5	5
Container (39 HD, 5 HT)	DCD	45				45
Ampoule	DCD		1			1
Subtotal		54	49	1	33	137
Total		68	56	17	33	174

[a]Does not include CAIS items.
[b]Determined not to be compatible with processing in TOCDF.
[c]6 HD, 2 HT.
[d]39 GB, 9 GD.

SOURCE: Provided to the committee by PMNSCM on July 10, 2001.

nitions currently stored at the four military sites in the United States that have the largest inventories of non-stockpile materiel.[14] According to the CWC, these recovered items must be destroyed by April 29, 2007. About 85 percent of all recovered CWM in the United States is stored at Pine Bluff Arsenal, in Arkansas (Table 1-1); smaller quantities are stored at Dugway Proving Ground, in Utah (Table 1-2), Aberdeen Proving Ground, in Maryland (Table 1-3) and Anniston Army Depot, in Alabama (Table 1-4). These four sites, as well as another three sites whose non-stockpile inventory totals 21 items, are addressed in further detail in Appendix D. Many more chemical munitions will be recovered at burial sites as current and former artillery ranges around the country are remediated. Whether the munitions recovered to date are representative of those that will be recovered in the future is an open question.

A brief description of each category of NSCWM follows.

Buried CWM

This is the most challenging and, at the same time, the most uncertain category. As noted, the Army has identified potential NSCWM burial sites in 31 states, the District of Columbia, and the U.S. Virgin Islands. In addition, potential overseas chemical weapons burial sites have been identified, but their locations are classified. Under the CWC, a state-party that has abandoned chemical weapons on the territory of another state-party has the responsibility for their removal and disposal (U.S. Army, 1996).

Burial sites are grouped in four categories: CAIS only; small quantity, nonexplosive (fewer than 1,000 items, agent

---

[14]Christopher Ross, PMNSCM, presentation to the committee on July 10, 2001.

TABLE 1-3 Inventory of Non-Stockpile Items at Aberdeen Proving Ground, Maryland

Item	Chemical Fill					Total
	HD/HT/HS	GB/GA/GD	VX	Lewisite	CG	
Munitions						
Explosive						
75-mm projectile	6					6
4.2-inch mortar	1				1	2
Nonexplosive						
4-inch Stokes mortar	2					2
Subtotal	9				1	10
Chemical sample containers						
55-gallon drum (pumpkins)	11					11
30-gallon bucket (pumpkins)	5	5	5			15
5-gallon bucket (steel cylinders)			20			20
5-pint can (vials or bottles)	3		23			26
Screw-top can (vials or bottles)			7			7
Steel cylinder		12				12
Multipack bottles, vials		6		10		16
DOT bottle			6			6
Ton container		2				2
Subtotal	19	25	61	10		115
Total	28	25	61	10	1	125

SOURCE: Provided to the committee by PMNSCM on July 10, 2001.

TABLE 1-4 Inventory of Non-Stockpile Items at Anniston Chemical Activity, Alabama

Item	Chemical Fill			Total
	HD/HT	GB	VX	
Chemical sample containers				
Vial		119		119
DOT bottle	5		7	12
Ton container		2		2
Total	5	121	7	133

SOURCE: Provided to the committee by PMNSCM on July 10, 2001.

present but no energetics); small quantity explosive (fewer than 1,000 items, agent and energetics); and large quantity (more than 1,000 items, including items configured both with and without energetics). Excluding sites at which no further action is required, estimates are that large-quantity sites make up 4 percent of all sites, small-quantity explosive sites make up 53 percent, small-quantity nonexplosive sites make up 32 percent, and CAIS-only sites make up 11 percent of the sites. The largest burial site is believed to be the old "O" field in Edgewood, Maryland; large sites also exist at Tooele Army Depot, Utah; Rocky Mountain Arsenal, Colorado; and Redstone Arsenal, Alabama (U.S. Army, 1996).[15]

A major uncertainty for the non-stockpile program is the extent to which suspected burial sites will be excavated and what items will be found and recovered. Remediation efforts are under way or planned at the following sites: former Camp

---

[15]No data are available on the percentages of the total non-stockpile inventory located at each type of site.

Sibert, Alabama; Fort McClellan, Alabama; former Santa Rosa Army Air Field, California; Spring Valley, Washington, D.C.; former Brooksville Army Air Field, Florida; England Air Force Base, Louisiana; Lauderick Creek, Aberdeen Proving Ground, Maryland; former Defense Depot, Memphis, Tennessee; Ogden Depot, Utah; former Lauringburg-Maxon Army Air Field, North Carolina; former Fort Segarra, U.S. Virgin Islands; and Camp Bullis, Texas. Remediation efforts at these sites are highly dependent on the availability of mobile treatment systems or sites that can receive and store the recovered NSCWM (or secondary wastes) prior to final disposal.

Remediation at other sites is expected to be planned and initiated. Because public stakeholder groups at potential recipient sites are very concerned about becoming "dumping grounds" for the nation's recovered NSCWM, the availability of such sites in the future remains uncertain (see Chapter 5). Since in any case the number of sites that can receive and store NSCWM prior to disposal is limited, a key solution is the development of transportable treatment systems that can be moved to a recovery site, used to treat the recovered chemical warfare materiel (RCWM), and then dismantled and moved to another site.

Binary Chemical Warfare Materiel Components

Binary weapons were developed in the 1980s to replace the aging stockpile of unitary chemical weapons. Two binary munitions were investigated, using two sets of agent precursors: the chemical QL, which could be reacted with powdered sulfur to produce the nerve agent VX, and the chemical DF, which could be reacted with a nonlethal solution, OPA (isopropyl alcohol with isopropylamine), to form the nerve agent GB (sarin). The only binary weapon system to reach full-scale production was the M-687 155-mm artillery projectile. The DF and OPA were stored separately until the projectile was to be fired, at which time a canister of DF was loaded into the round already containing a canister of OPA. The shock of firing the round ruptured the barrier between the chemicals, while the spin mixed the chemicals and facilitated their reaction on the way to the target.

The M-687 projectiles and the OPA canisters were stored at Umatilla Chemical Depot, in Oregon, and they had all been destroyed by March 1999 (U.S. Army, 1999a). As indicated in Table 1-1, almost 57,000 canisters and 7 drums of DF, as well as 293 drums of QL, are still stored at Pine Bluff Arsenal, Arkansas.[16] Miscellaneous quantities of DF and QL stored at Aberdeen Proving Ground have been destroyed.[17]

Recovered Chemical Warfare Materiel

CWM items retrieved from range-clearing operations, research and test sites, and burial sites are classified as recovered chemical warfare materiel (RCWM).[18] Of the more than 7,000 RCWM items recovered to date, more than 5,400 are CAIS sets or CAIS components. Because CAIS items were used for training for defense against chemical attack, the CWC does not require their disposal; however, the Army has decided to destroy them along with the other non-stockpile items (NRC, 1999a). Virtually all recovered NSCWM in the United States is stored at stockpile sites.[19] The only exceptions are moderate quantities of CAIS stored in places such as Fort Richardson, Alaska, and Camp Bullis, Texas. This coincidence naturally suggests the possibility of destroying the stored RCWM in stockpile disposal facilities. Most of the remaining recovered non-stockpile munitions (about 1,250)—by far the most difficult type of RCWM to dispose of—are stored at Pine Bluff, Arkansas (see Table 1-1).[20] These munitions are currently being analyzed to determine the type of agent and energetics contained. Small numbers of non-stockpile munitions, including RCWM, are also stored at Dugway Proving Ground, Utah (37 items) (Table 1-2) and Aberdeen Proving Ground, Maryland (10 items) (Table 1-3).

Former CWM Production Facilities

The CWC requires that all former CWM production facilities constructed or used after January 1, 1946, be destroyed. The United States has declared 13 former production facilities in seven states under the CWC, although NSCMP does not have exclusive responsibility for destroying all of these (U.S. Army, 1996). NSCMP has made substantial progress in destroying the facilities for which it is responsible.

Miscellaneous CWM

Miscellaneous CWM includes the following:

• items designed specifically for conducting chemical warfare, such as unfilled munitions, empty rocket warheads, fuzes and bursters designed for chemical munitions, and simulant-filled munitions

• chemical samples transferred from leaking or suspect munitions to safer storage containers

---

[16]Christopher Ross, PMNSCM, presentation to the committee on July 10, 2001.

[17]Bill Brankowitz, Office of the PMNSCM, presentation to the committee on January 22, 2001.

[18]RCWM is defined by Army Regulation (AR) 50-6. See Chapter 5 for details.

[19]Stored RCWM also includes chemical samples, see below.

[20]RCWM recovered from other sites (e.g., the Spring Valley site in Washington, D.C.) has, in most cases, been shipped to Pine Bluff Arsenal for storage.

- ton containers, now empty, in which chemical warfare agents were previously stored or shipped
- research, development, test, and evaluation (RDT&E) CWM used for the development of CW

The CWC requires disposal of the first two groups of miscellaneous materiel (Blackwood, 1998). The ton containers are not controlled by the CWC because they are used to store commercial chemicals as well as chemical warfare agents. While the Army currently lists no items in the RDT&E group, it may reclassify some chemical sample materiel as RDT&E for future chemical defense research allowed by the CWC (Blackwood, 1998).

The Army appears to be making good progress in destroying unfilled chemical munition components and empty ton containers. For this reason, the only type of miscellaneous CWM considered in this report is the chemical samples. These are stored at stockpile sites in a wide variety of containers, from ton containers to glass vials; they contain a variety of chemical agents.

Systems for Destroying NSCWM

The Army has developed a mix of semi-permanent[21] facilities and mobile systems to meet its obligations under the CWC to destroy all recovered NSCWM by 2007. Before the systems can be operated, however, the Army must establish their technical effectiveness and safety, obtain the necessary regulatory approvals for operation at each site where NSCWM is located, and provide opportunities for the various public stakeholder groups to be involved in the decision-making process.

Facilities

As noted earlier, Congress has relaxed its former prohibition on the use of stockpile facilities to destroy NSCWM, provided that their use is acceptable to the states in which the facilities are located. These facilities will be in operation until at least 2007 and may now be considered as an option for destroying NSCWM that is co-located at the facilities or that can be safely transported to them.

In addition to the stockpile facilities, two experimental facilities have long been used to destroy a variety of chemical agents by chemical neutralization: the Chemical Transfer Facility (CTF) at Aberdeen Proving Ground, Maryland, and the Chemical Agent Munitions Disposal System (CAMDS), at the Deseret Chemical Depot, Utah. Although these are R&D facilities and therefore may not be used on a routine basis to destroy NSCWM, they might be considered as an option to destroy limited numbers of non-stockpile items that contain unusual chemical fills or have a configuration that cannot be handled by other systems.

At some sites where large numbers (hundreds or thousands) of NSCWM items are stored or expected to be recovered, the Army plans to build facilities that will have a higher processing capacity than mobile systems. At this writing, construction of one such facility, the Munitions Assessment and Processing System (MAPS) at Aberdeen Proving Ground, was under way. A second, the Pine Bluff Non-Stockpile Facility (PBNSF) at Pine Bluff Arsenal, Arkansas, was entering its final design phase. Each of these facilities will use an array of technologies to destroy the non-stockpile chemical agents.

Mobile Destruction Systems

Because a large number of locations are expected to have only a small quantity of NSCWM, the Army has been particularly interested in developing transportable disposal systems that can be taken from site to site as needed. To treat the diversity of NSCWM (e.g., munitions containing a variety of chemical agents, some configured with explosives and some without), the Army possesses two transportable systems, the Rapid Response System (RRS) and the Explosive Destruction System (EDS). The RRS is designed to treat chemical agent identification sets (CAIS), test kits used from 1928 to 1969 to train soldiers to identify chemical agents in the field. CAIS contain mustard and lewisite, as well as a variety of toxic industrial chemicals[22] (NRC, 1999a). The EDS is designed to destroy explosively configured nonstockpile munitions up to 155-mm projectiles or 8-inch World War I chemical projectiles. Small numbers of nonexplosively configured munitions or containers could also be destroyed in the EDS, although larger numbers or larger sizes of such containers would be expected to be processed in the facilities discussed above—that is, MAPS and PBNSF.

These systems, which can be mounted on a series of trailers, are designed to be transported to a site where the chemical materiel is located and then packed up and moved to another site upon completion of the treatment campaign. They use chemical neutralization processes to treat chemical agents; secondary waste from the neutralization process can

---

[21]The U.S. Army Corps of Engineers defines semi-permanent facilities as having a life expectancy of from 5 to 25 years. For the purposes of this report, the term "facilities" refers to semi-permanent facilities.

[22]In keeping with Army practice, all chemicals used as CWMs other than the nerve agents (e.g., VX and GB) and blister agents (e.g., HD and L) are referred to as "toxic industrial chemicals." This includes choking agents (e.g., phosgene or CG), vomiting agents (e.g., adamsite, or DM), and lachrymators (e.g., chloropicrin (CS) and choroacetophenone (CN)). In most instances in this report, the term "chemical agents" is restricted to the nerve and blister agents.

be treated further on-site or shipped to an off-site treatment facility.

STATEMENT OF TASK

The Army has available, or is in the process of developing, a number of alternative systems for addressing the many contingencies it faces in its task of destroying the diverse categories of NSCWM located at many different sites. To date, however, there has been little integrated review of the technical operational plans for these systems—how they fit together as an integrated toolbox and how their capabilities match the task to be accomplished. Equally important, there has been little review of the Army's plans for obtaining regulatory approvals to use these systems in the many states in which it must operate, and for providing opportunities for public involvement in the decision-making process so that the NSCWM can be destroyed in a timely way.

Accordingly, on March 16, 2001, the PMNSCM requested that the National Research Council undertake such a review. The statement of task is as follows:

> To help optimize the technical performance, as well as the regulatory approval and public acceptance processes, of the Non-stockpile Chemical Materiel Disposal program, the NRC will:
>
> • Evaluate mobile destruction systems and semi-permanent facilities being used or considered by the Army's Non-stockpile product manager for the treatment of non-stockpile CWM and make recommendations on the systems and facilities that could be employed by the Army and their interrelationships. This analysis will specifically include consideration of issues and opportunities associated with the Explosive Destruction System (EDS), the Rapid Response System (RRS), the Munitions Assessment and Processing System (MAPS), the Pine Bluff Non-stockpile Facility (PBNSF), alternative treatments for neat chemicals, and selected aspects of the stockpile facilities.
> • Review and evaluate the issues and obstacles associated with the environmental regulatory approval process for successful employment of Non-stockpile Chemical Materiel disposal systems (mobile and semi-permanent) that the Army may encounter during its management of the Non-stockpile Program and offer recommendations that may make the regulatory approval process more efficient while reducing schedule risk.
> • Recommend areas in which further detailed study efforts would be particularly useful to the Product Manager.

THE COMMITTEE'S APPROACH

The committee focused its attention primarily on the mobile treatment systems (the EDS and the RRS) and the individual treatment technologies that will eventually be components of the MAPS and PBNSF, and on NSCMP's operational plans for deploying these facilities and systems to destroy recovered NSCWM before April 29, 2007, in accordance with the CWC. The potential role of the stockpile facilities in destroying NSCWM was also reviewed. The committee paid particular attention to technical issues but also considered operational plans and schedules in light of the regulatory and public involvement challenges that they will face. The Army has prepared assessments of the potential health and environmental impacts of its transportable treatment systems (U.S. Army, 2001a), including risks during normal operations and from accidental release of hazardous substances. Similar site-specific assessments are generally required of the Army's treatment facilities as part of the permitting process. The committee did not review the specific methodology or the regulatory assumptions used by the Army in assessing these health and environmental impacts, because the overall risk assessment methodologies are of the kind typically used in U.S. regulatory and permitting programs.

SCOPE OF THE REPORT

The Army appears to be making excellent progress in destroying several categories of NSCWM, such as old production facilities, empty ton containers, and unfilled CWM delivery systems. For this reason, these categories of NSCWM are not discussed in this report, which instead focuses on the several subcategories of NSCWM whose destruction poses the greatest challenges: chemical munitions, binary CWM components, chemical samples, and CAIS. Destruction of the secondary waste streams generated in the treatment of these NSCWM is considered along with the primary treatment.

STRUCTURE OF THE REPORT

This chapter described the non-stockpile inventory, which comprises a greater variety of chemical agents than the stockpile inventory. It includes blister, nerve, blood, and choking agents, as well as militarized industrial chemicals and binary agents, and its condition is highly variable—some items, for instance, have severely deteriorated during decades of burial.

In Chapter 2, the committee assesses the tools, or specific options, available to PMNSCM to safely destroy these items. These tools, which include facilities, mobile treatment systems, and individual treatment technologies, are evaluated from the standpoint of their current status and technical, RAP, and public involvement issues.

In Chapter 3, the committee matches the treatment options with the materiel or munitions to be treated, identifies gaps in the program, and makes recommendations on the facilities, systems, and technologies. Chapter 4 examines RAP issues for waste management and identifies issues that, when resolved, will facilitate the RAP process in the future. Chapter 5 commends PMNSCM for its increased openness in providing information to a range of stakeholders and in developing relationships with them and notes areas that might be improved. Throughout the report, findings and recommendations follow the relevant discussion.

2

The Toolbox of Non-Stockpile Treatment Options

The Army has a number of options available for the treatment of NSCWM. They include the use of facilities designed to treat both non-stockpile and stockpile CWM, the use of mobile systems that can be incorporated into a facility or transported to the site of a find, and individual treatment technologies. Like mobile systems, individual treatment technologies may be incorporated into a larger entity such as a facility or mobile system or transported to the site of a find. This chapter examines these options, or tools, and, where appropriate, presents the committee's findings and recommendations related to their use. Each facility, mobile system, and individual technology is examined from the standpoint of its current status, as well as technical, RAP, and public involvement issues. An overview of the options for treating NSCWM considered in this chapter appears in Table 2-1.

The discussion in this chapter is based on information the committee was able to gather on available test results or operational plans for use of the tools as presented by the Army. However, equipment does not always function as designed, and unexpected events—even catastrophic failures—may occur. The Army has conducted preliminary accident risk assessments of treatment systems (see, for example, U.S. Army, 2001a, Appendix D, "Summary of Accident Risk Assessment") but does not appear to have conducted an integrated site-specific risk assessment of its systems, including the risks of catastrophic failures.

As discussed in Chapter 1, non-stockpile sites span a considerable range—from sites at which large numbers of non-stockpile munitions are buried or stored, to sites containing only a few chemical agent identification set (CAIS) vials or bottles. In some instances, the non-stockpile sites are at current military facilities, where the Army has full control and consequently has several treatment options from which to choose; in other cases, the sites are at former defense facilities that are now commercial or residential properties, where treatment options may be more limited.

Some treatment options, such as the use of stockpile incinerators, would destroy the non-stockpile item directly. Others, especially those involving chemical neutralization, generate liquid secondary waste streams that require further treatment before disposal. This secondary waste treatment could take place in a commercial treatment, storage, and disposal facility (TSDF) or could employ one or more of the individual alternative technologies, such as chemical oxidation, either at the site where chemical neutralization takes place or at an off-site location. If secondary waste is defined as hazardous waste, such treatment would need to be conducted at a commercial TSDF permitted or approved by the appropriate regulatory authority under the Resource Conservation and Recovery Act (RCRA).

NON-STOCKPILE FACILITIES

The Army has at its disposition four principle types of facilities for treating non-stockpile chemical materiel: non-stockpile facilities, designed to destroy large quantities of dissimilar CWM; stockpile facilities, constructed to destroy large quantities of similar CWM; research and development facilities; and commercial treatment, storage, and disposal facilities (TSDFs).

TABLE 2-1 Overview of Non-Stockpile Treatment Options

Treatment Option	Description
Facilities	
Non-stockpile facilities	
Pine Bluff Non-Stockpile Facility (PBNSF) (in final design)	Designed to use chemical neutralization and associated technologies to address the recovered non-stockpile items stored at Pine Bluff Arsenal, Arkansas
Munitions Assessment and Processing System (MAPS) (under construction)	Designed to use chemical neutralization and associated technologies to address the recovered non-stockpile items found at Aberdeen Proving Ground, Maryland
Use of stockpile destruction facilities for disposal of non-stockpile materiel	Equipped to open stockpile chemical munitions, drain and incinerate agent, and destroy energetics
Research and development facilities	
Chemical Transfer Facility (CTF)	Research facility at Aberdeen Proving Ground, Maryland, capable of destroying stockpile and non-stockpile agents
Chemical Agent Munitions Disposal System (CAMDS)	Research facility at Tooele, Utah, capable of destroying non-stockpile munitions containing agent fills not easily accommodated at other facilities, e.g., lewisite
Treatment, storage and disposal facilities (TSDFs)	Capable of high-temperature incineration of secondary waste streams produced by the RRS, EDS, and other systems
Mobile Treatment Systems	
Rapid Response System (RRS)	Mobile trailer system to handle numerous CAIS vials and/or PIGs found in one location
Single CAIS Accessing and Neutralization System (SCANS) (in design)	Small reactor in which individual CAIS vials or bottles can be crushed and neutralized
Explosive Destruction System (EDS)	Mobile trailer system in which explosively configured munitions are explosively accessed and their chemical contents are neutralized
Donovan blast chamber (DBC) (in testing for use with CWM)	Mobile system potentially usable for the destruction of explosively configured munitions without neutralization of their chemical contents
Individual Treatment Technologies	
Plasma arc	High-temperature technology for direct destruction of agent or for destruction of secondary waste streams produced by the RRS, EDS, and other systems
Chemical oxidation	Low-temperature technology potentially applicable to destruction of liquid secondary waste streams produced by the RRS, EDS, and other systems
Wet air oxidation	Moderate-temperature technology potentially applicable to the destruction of liquid secondary waste streams produced by the RRS, EDS, and other systems
Batch supercritical water oxidation (SCWO)	High-temperature technology still at the R&D stage that is potentially applicable to destruction of neat agent and CAIS vials
Neutralization (chemical hydrolysis)	Low-temperature technology for hydrolysis of neat chemical agents and binary precursors
Open burning/open detonation (OB/OD)	Historic blow-in-place method for destroying dangerous munitions
Tent and foam	Partially contained blow-in-place method for destroying dangerous munitions

MAPS and PBNSF

Unlike stockpile facilities, discussed next, no dedicated non-stockpile facilities have yet been completed. The PMNSCM plans to construct such facilities at two sites where large quantities of recovered chemical warfare material (RCWM) are stored: the Munitions Assessment and Processing System (MAPS) at Aberdeen Proving Ground (APG), Maryland, and the Pine Bluff Non-Stockpile Facility (PBNSF) at Pine Bluff Arsenal (PBA), Arkansas.

Both MAPS and PBNSF are designed to treat non-stockpile chemical agents by an array of neutralization technologies, although future facilities might be based on other treatment methods. The neutralization processes to be used at MAPS and PBNSF are based on those developed for the now-defunct Munitions Management Device (MMD) systems, involving aqueous monoethanolamine (MEA) for HD and GB and aqueous caustic for CG (phosgene) (NRC, 2001a). Arsenic-containing agents such as lewisite would probably be treated with sodium hydroxide (NRC, 2001b).

MAPS and PBNSF will also have components for unpacking and characterization of NSCWM, mechanical accessing of the chemical agent in munitions or containers, and explosive destruction of energetics. Secondary wastes from the neutralization process may be destroyed on-site or shipped off-site for treatment.

This evaluation focuses on the current state of planning and construction of MAPS and PBNSF. For this reason, issues such as decommissioning and decontamination are not considered here.

Munitions Assessment and Processing System

The MAPS facility, discussed in detail in Appendix D, has been designed to deal with explosively configured chemical munitions and smoke rounds that will be recovered during the Installation Restoration Program (IRP) at APG. APG has been used for testing chemical weapons for more than 70 years, and the types and numbers of items that will be recovered are unknown at this time.

A floor plan for MAPS is shown in Figure 2-1. Operators will drill or cut the munition and drain the chemical agent from the munition body in an explosive containment chamber. The separated explosives will then be detonated in a commercial detonation vessel. The chemical agent will be transported to and neutralized in the Chemical Transfer Facility (CTF) already located at APG.

MAPS is designed to process munitions as large as a 155-mm projectile. MAPS could treat a maximum of seven to eight munitions per day, depending on the agent and the condition of the munitions.

Pine Bluff Non-Stockpile Facility

The PBNSF, also discussed in detail in Appendix D, will be designed specifically to process RCWM, binary chemical weapons components, CAIS, and chemical samples at PBA. The present inventory (Table 1-1) lists 69,878 items, including explosive and nonexplosive munitions, chemical sample containers, CAIS, binary CWM precursors, and empty ton containers. The chemical fills are diverse and include chemical agents and their mixtures (H, HS, HD, HN3, L, HL, H-CHCl$_3$, L-CHCl$_3$), industrial chemicals (CG, DM, CK, PS, PS-CHCl$_3$), and binary precursors (DF and QL). The nerve agent GB is contained in two partially filled ton containers and in research, development, test, and evaluation (RDT&E) vials, and VX is in RDT&E containers.

Of the 69,878 non-stockpile items at Pine Bluff Arsenal, the two GB-filled ton containers, possibly the nine M70A1 bombs, two chemical containers filled with HD and VX, and 5,814 mustard-filled CAIS could be processed in its stockpile facility. The remaining items are more suitable for processing in the PBNSF or in mobile systems such as the RRS or EDS.

Stable munitions will be placed in the explosive containment chamber (ECC), where they will be drilled and drained, the chemical agent neutralized, and the liquid neutralents collected. Following the drill-and-drain operation, the munition will be pressure-rinsed with reagent and water, bagged, and transferred to the detonation chamber, where the energetics will be deactivated using auxiliary explosives. The resulting shrapnel fragments will be monitored for residual agent to verify that they are agent free. Liquid neutralents will be sent to a secondary treatment facility, whose process technologies have yet to be specified. Metal parts, when decontaminated to a level 3X,[1] will be disposed of in an appropriately permitted facility.

RCWM such as projectiles and mortar rounds could be processed in various versions of the EDS or by drilling and draining in an explosion containment room at PBNSF. Munitions that are designated unstable will be processed in an EDS outside the facility. CAIS items and chemical sample bottles could be handled in an RRS or, as currently planned, in glove boxes and neutralization reactors in the PBNSF, with the neutralent sent to a TSDF. Munitions not containing energetics will be destroyed in a fashion similar to that used for the explosively configured items, except that after the agent has been drained, the munition will be sent to a cutting station, where it will be reduced in size.

---

[1] 3X refers to the level at which solids are decontaminated to the point that agent concentration in the headspace above the encapsulated solid does not exceed the health-based, 8-hour, time-weighted average limit for worker exposure. The level for mustard agent is 3.0 mg/m^3 in air. Materials classified as 3X may be handled by qualified plant workers using appropriate procedures but are not releasable to the environment or for general public reuse. In specific cases in which approval has been granted, a 3X material may be shipped to an approved hazardous waste treatment facility for disposal in a landfill or for further treatment.

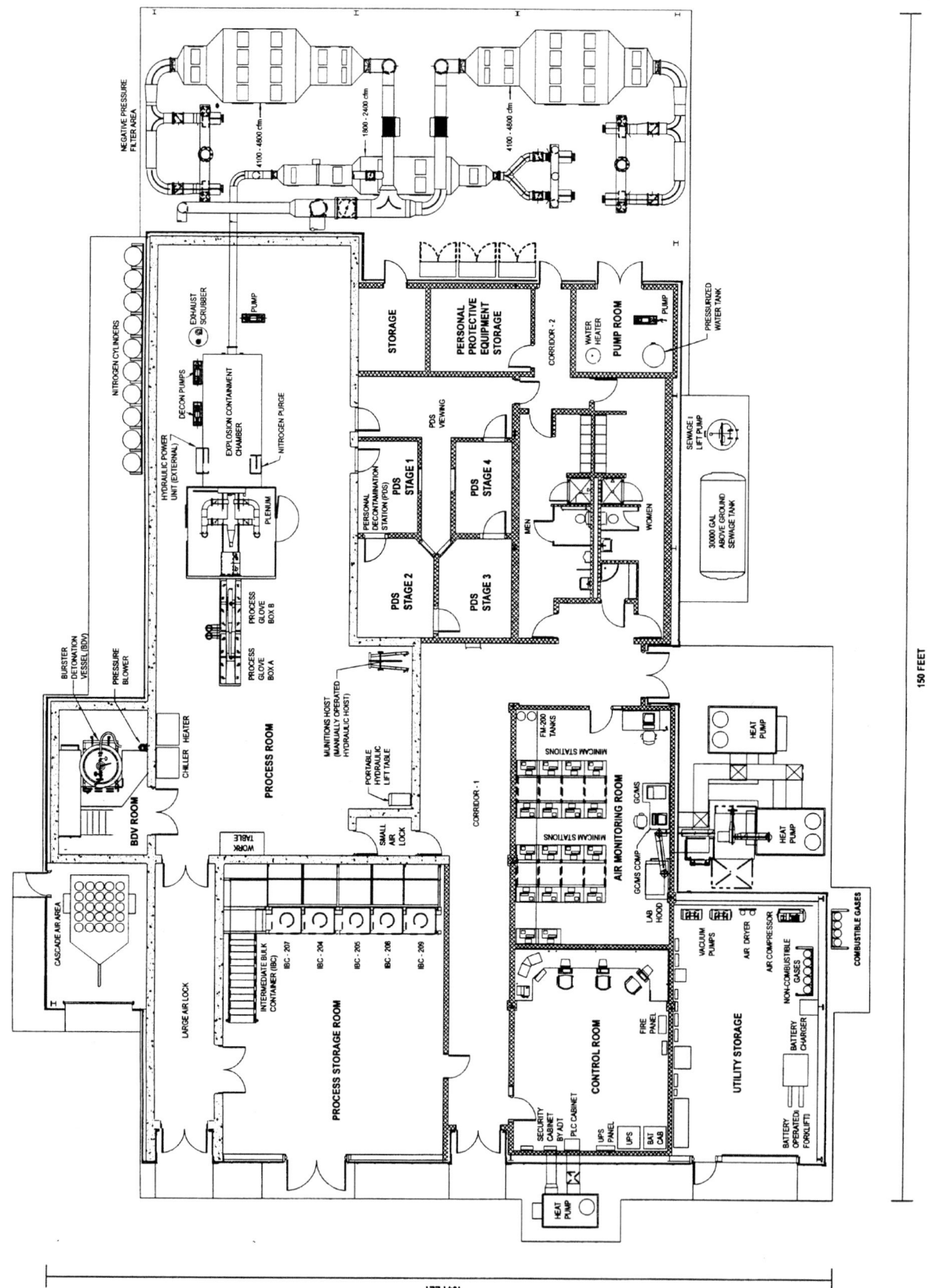

FIGURE 2-1 Floor plan of MAPS. SOURCE: Provided to the committee by Don Benton, Office of the PMNSCM, July 12, 2001.

Binary CWM components could be processed by destroying the QL and DF by supercritical water oxidation (SCWO), plasma arc, or another nonincineration technology. At this writing, it has not been decided how the binary CWM components at PBA will be processed.

Status of MAPS and PBNSF

There are currently no functioning non-stockpile facilities. The design of the MAPS is completed, and construction has begun. The Army's plan is to conduct a facility demonstration during FY 2002-2004, with actual disposal running from FY 2004 to 2009. The design of the PBNSF is being finalized, with construction scheduled to begin in January 2004. PMNSCM plans that PBNSF will operate from April 2006 to April 2007, completing destruction of all items on hand in time to meet the CWC deadline of April 29, 2007. In addition to incineration, two technologies are currently being evaluated for secondary waste treatment: supercritical water oxidation and plasma arc.

Technical Issues

Relatively few non-stockpile munitions are stored at APG (Table 1-3), although large numbers are believed to be buried there. The maximum anticipated throughput rate of eight munitions per day should be adequate to destroy the stored non-stockpile munitions as well as any recovered non-stockpile munitions at APG in a reasonable time.

By contrast, the non-stockpile inventory at PBA has large numbers of non-stockpile munitions and containers of agent in storage and relatively few buried munitions. Because the PBNSF facility is still being designed, there are no reliable estimates of the throughput rate for destruction of the various types of NSCWM located there. Based on the schedule PMNSCM provided the committee (U.S. Army, 2001c), much of the PBA inventory is to be destroyed in 2006 and 2007. However, the committee finds the PMNSCM destruction schedule to be overoptimistic and concludes that the CWC deadline of April 29, 2007, will almost certainly not be met.

At this writing, two technologies are being considered for treatment of secondary liquid wastes generated at PBNSF: SCWO and plasma arc. Both of these technologies face technical and/or permitting challenges that could add to concerns about meeting the CWC deadline for destruction of NSCWM at PBA. The SCWO technology, for example, was recently tested in the Assembled Chemical Weapons Assessment (ACWA) program and found to have operational problems that may make it unattractive for the disposal of certain chemical wastes. The SCWO reaction is so corrosive that it erodes the reactor container, in response to which sacrificial liners are inserted in the reactor. The choice of materials of construction for these liners is dependent on the elemental composition of the feedstock (feedstocks containing chlorine and fluorine are particularly corrosive). Indeed, the engineering design studies for the ACWA program indicate that in many cases the protective liner in the reactor must be changed every few weeks (NRC, 2002b). Also, the salts produced are insoluble in supercritical water and plug the reactor, requiring frequent shutdown and flushing.

Plasma arc technology has been used successfully in Europe to destroy chemical warfare material but has not been permitted in the United States. Currently, PMCD is optimistic that it will have little difficulty in obtaining a permit. They have identified several plasma arc firms in this country that have operational units, but none has destroyed a CW-related waste stream. If the ACWA program does not develop a continuous SCWO system that is cost-effective for use on the quantities of materiel to be destroyed in the non-stockpile program and if a permit for the plasma arc technology cannot be obtained in time, the Army may be forced to incinerate its waste streams to comply with the CWC treaty deadline of April 2007.[2]

Regulatory Approval and Permitting Issues

MAPS will operate initially under a Resource Conservation and Recovery Act (RCRA) Research, Development, and Demonstration (RD&D) permit and will transition to a standard RCRA Part B permit when operations become routine. The Army has worked closely with Maryland regulators, and the MAPS permitting process has gone relatively smoothly.

The Army has just begun to evaluate permitting strategies for the PBNSF. As with MAPS, construction and initial operation of the facility could proceed under a RCRA RD&D permit. After operations at the facility become routine, the PBNSF operations could be transitioned to the full RCRA permit. Permitting activities for PBNSF are still at an early stage, and with the final treatment technology yet to be decided, it remains unclear what problems may be encountered.

Public Concerns

As noted above, semi-permanent facilities are designed specifically to allay public concerns that the facilities will remain open permanently and become "magnets" for off-site wastes. Using chemical neutralization rather than incineration to destroy chemical agents may also alleviate public concerns about potential hazardous air emissions.

In the case of MAPS, the Army appears to have worked diligently to gain the confidence of local public interest

---

[2]While some interest groups have advocated storage of neutralent waste streams until a viable technology alternative to incineration can be developed, the Army has indicated that it would not consider long-term storage of secondary wastes at PBA.

groups. The Army's public outreach efforts for PBNSF are at a very early stage (see Chapter 5).

Finding 2-1a. PBA has the largest known non-stockpile inventory. It contains almost 70,000 items, including explosive and nonexplosive munitions with diverse chemical fills, binary agent precursors, CAIS, chemical samples, and empty ton containers. If the CWC treaty deadline is not extended, these items must be destroyed by April 29, 2007.

Finding 2-1b. The task of destroying the very large quantity of NSCWM at Pine Bluff Arsenal by 2007 is daunting, given that the planned non-stockpile facilities are not expected to be operational until 2006. As far as the committee can ascertain, the Army has not developed a timetable for destruction of this NSCWM that is both realistic and consistent with current treaty deadlines. The committee is concerned that without clear planning and extraordinary efforts, the treaty deadlines will almost certainly not be met.

Recommendation 2-1. PMNSCM should develop a detailed, realistic timetable showing how the planned non-stockpile facilities at Pine Bluff Arsenal can achieve the throughput necessary to destroy the stored non-stockpile items by April 2007 and should communicate this timetable to all stakeholders.

Finding 2-2. The construction of semi-permanent facilities is a valid approach at sites where large quantities of NSCWM are stored or expected to be recovered. Installing equipment in buildings instead of trailers, tents, and other temporary facilities will provide a more comfortable and safer environment for the workers. Also, more thorough environmental safeguards can be provided than in mobile facilities, thus lowering the risk of fugitive gases escaping into the environment.

Recommendation 2-2. The Army should consider constructing treatment facilities at other sites where large quantities of NSCWM are expected to be recovered.

Stockpile Facilities

Until November 1999, federal law prohibited the use of stockpile chemical disposal facilities for destroying anything other than stockpile CWM. In P.L. 106-65, Congress amended the law to allow non-stockpile materiel to be destroyed in stockpile facilities, provided the states in which the stockpile facilities are located agree, thus creating a new treatment option for non-stockpile materiel.

The chemical stockpile contains projectiles, mortar rounds, rockets, land mines, bombs, spray tanks, and bulk containers that are filled with the blister agent mustard and the nerve agents GB and VX. Under the Chemical Stockpile Disposal Program, the Army is in the process of planning, constructing, and operating chemical disposal facilities (CDFs) at eight locations. Appendix C evaluates the suitability of stockpile chemical disposal facilities for treating non-stockpile CWM that is co-located at these facilities.

Description

At four of the eight storage locations in the United States (Tooele Chemical Disposal Facility, Utah [TOCDF]; Anniston Chemical Activity, Alabama; Umatilla Chemical Depot, Oregon; and Pine Bluff Arsenal, Arkansas), the agent, energetics, and agent-contaminated metal surfaces will be processed in appropriately designed incinerators and furnaces. Demilitarization machines will be used to access the agent through techniques such as reverse assembly of projectiles and mortar rounds, punching and draining of bulk containers and land mines, and draining and shearing of rockets. The CDF incinerators, pollution abatement systems, monitoring equipment, and carbon filtration systems are designed to process the agent fills found in the stockpile munitions and bulk containers.

Status

Agent operations with the baseline incineration system have been completed in the Johnston Atoll Chemical Disposal System (JACADS) and closure operations have begun (U.S. Army, 2000b; NRC, 2002a). Operations continue at TOCDF, which has processed almost two-fifths of its stockpile. Construction has been completed at Anniston, Alabama (Anniston Chemical Disposal Facility), and Umatilla, Oregon (Umatilla Chemical Disposal Facility), where systemization (operational testing) is under way. Construction at Pine Bluff, Arkansas (Pine Bluff Chemical Disposal Facility) is approximately two-thirds complete. A number of chemical disposal facility concepts, including incineration and alternative technologies, are in the planning and design stage as competing options for Pueblo Chemical Depot, Colorado (Pueblo Chemical Disposal Facility—PUCDF) (NRC, 2001c). Three alternative technology options are currently under consideration by the ACWA program for use at Blue Grass Army Depot, Kentucky.

Recently, the Army concluded that several of the CDF incinerators will remain in operation beyond April 29, 2007, the original goal for completion of the destruction of the chemical stockpile under the CWC (U.S. Army 2001c). PMNSCM has recommended in several cases that small, non-stockpile chemical sample containers currently in storage at stockpile incinerator locations (e.g., Department of Transportation (DOT) approved bottles, vials, drums, steel cylinders, and ampoules having VX, GB, and mustard fills) should be processed in the CDFs.

Technical Issues

The CDFs are specifically designed to process stockpile items and agent fills (GB, VX, and mustard). The measured destruction and removal efficiencies of the stockpile incinerators exceed 99.9999 percent and include stringent controls on emissions of hazardous air pollutants (Smithson, 1994).

In Appendix C, the technical feasibility of destroying non-stockpile items in stockpile facilities is evaluated as an alternative or supplement to disposing of these items in either other (non-CDF) facilities or non-stockpile mobile equipment such as the RRS and EDS.

Incinerators for destruction of stockpile materiel are or will be available at Umatilla, Tooele, Anniston, and Pine Bluff. The suitability of the incinerators and associated infrastructure for destroying non-stockpile items is as follows:

- *Umatilla.* The existing permit allows destruction of NSCWM, and the state of Oregon prefers to use the stockpile chemical disposal facility (CDF) for destruction of the five non-stockpile ton containers stored at Umatilla.
- *Deseret.* The incineration facility at Tooele provides a technically feasible alternative for destruction of 157 of the 174 non-stockpile items stored at Deseret. The other 17 items contain lewisite and are better suited for destruction at the Chemical Agent Munitions Disposal System (CAMDS). The Tooele CDF is scheduled to complete its stockpile mission in the fourth quarter of 2003, so it would be available to treat the non-stockpile items. Existing permits would need to be modified and the public would need to agree.
- *Anniston.* The CDF at Anniston provides a technically feasible alternative for destruction of the 133 non-stockpile chemical sample containers stored there. Permits would need to be obtained and the public would need to agree.
- *Pine Bluff.* The incineration facility at Pine Bluff provides a technically feasible alternative for destruction of less than 10 percent of the 69,878 non-stockpile items stored there because its design does not include facilities for opening bulk containers of agent and CWM binary components (Appendix C). However, inclusion of such capability in the non-stockpile PBNSF would enable transfer of these liquid chemicals to vessels suitable as feed tanks for the PBCDF liquid incinerator. This modification, plus the addition of DF and QL monitoring systems at the Pine Bluff Chemical Disposal Facility (PBCDF), would allow incineration of the great majority of the PBA non-stockpile inventory.
- The four stockpile incineration facilities might be able to secure public acceptance for treatment of non-stockpile materiel stored at other locations within the state, but even that is likely to be difficult. Acceptance of non-stockpile materiel from outside the state is extremely unlikely.
- Because facilities for the destruction of the stockpile at Pueblo and Blue Grass have not yet been selected, the committee makes only general comments on their suitability for destroying non-stockpile materiel (Appendix C).

Aberdeen has selected neutralization followed by off-site posttreatment at a commercial TSDF for destruction of its stockpile materiel. That system could treat at most 19 mustard-filled chemical sample containers of the 125 non-stockpile items stored at Aberdeen. MAPS is intended to treat the remaining 106 currently known non-stockpile items, as well as materiel recovered during the installation restoration program at APG.

Technically, it should also be possible to use CDFs to destroy secondary non-stockpile waste streams, whether they are generated from non-stockpile treatment processes conducted at stockpile sites or remote locations.

Regulatory Approval and Permitting Issues

Public Law 106-65 provides that non-stockpile materiel may be disposed of in stockpile facilities if the state in which the facility is located issues the appropriate permit or permits. However, in many stockpile locations, the original CDF permits exclude or severely restrict the destruction of any materiel other than stockpile materiel at these facilities. For example, at Pine Bluff Arsenal, Arkansas, the RCRA permit for the stockpile incinerator under construction prohibits the processing of any hazardous waste, including non-stockpile materiel, in the facility, whether the waste is located at the site or off-site (Arkansas, 1999). At Umatilla Chemical Depot, in Hermiston, Oregon, the stockpile facility's RCRA permit requires that the small quantity of non-stockpile materiel currently stored at the site be destroyed at the stockpile facility, but it prohibits any off-site hazardous wastes to be brought, stored, or treated at the facility (Oregon, 1997). Thus, permitting can represent a significant obstacle to the use of stockpile facilities to destroy NSCWM at most sites.

Public Concerns

The histories of the stockpile program, the non-stockpile program, and the use of mobile incinerators in hazardous waste site cleanups demonstrate that classical incineration often generates strong public opposition. There are concerns over the impact on local communities of the potential emission of small amounts of chemical agents not destroyed, as well as low concentrations of chemicals that are inevitably formed inside incinerators (e.g., polychlorinated dibenzodioxins) (Greenpeace, 2001; EPA, 1998; Sierra Club, 2001).

In the case of stockpile incinerators, public concerns at some sites have focused on the possibility that the facilities would continue to operate indefinitely after the stockpile was destroyed and would import hazardous wastes. If CDF permits were to be modified now to allow the destruction of non-stockpile items, some public stakeholders would probably feel that the Army had violated the commitments it made when it applied for the original permits for CDF operations. On the other hand, at some sites the public may feel that use of the CDFs to destroy non-stockpile items provides a cost-effective and expeditious means of eliminating the risk associated with the storage of its non-stockpile items.

Finding 2-3. The stockpile chemical disposal facilities (CDFs) are technically capable of safely disposing of a portion of the non-stockpile inventory, including secondary wastes, but could face challenging regulatory and public acceptability hurdles, especially if they accepted NSCWM from other sites or out of state. Although any NSCWM or secondary waste might be a candidate, materiel already located at stockpile sites—or, secondarily, located within the same state—may present less of an acceptance problem than NSCWM from other states.

Recommendation 2-3. Provided regulatory and public acceptability issues can be resolved, any NSCM located at stockpile sites and suitable for destruction in chemical stockpile disposal facilities should be destroyed in those facilities. This recommendation applies to non-stockpile materiel, secondary wastes, binary CWM components, and bulk chemicals.

Research and Development Facilities

The Army also has at its disposition two R&D facilities that might be used for treating appropriate non-stockpile items. PMNSCM has proposed using the Chemical Transfer Facility (CTF) at Aberdeen Proving Ground (APG) to destroy CWM recovered at APG and the Chemical Agent Munitions Disposal System (CAMDS), at Deseret Chemical Depot, to destroy non-stockpile items containing the arsenical agent lewisite.

Chemical Transfer Facility

The Chemical Transfer Facility (CTF) is an R&D facility at APG that has processed munitions, sample bottles, and ton containers containing a variety of chemical fills. The CTF is not capable of processing explosively configured munitions but does contain a chemical agent transfer system that can drain ton containers. There is no treaty-imposed time limit on operation of the CTF, and if its schedule permits, it can dispose of the container items listed above. PMNSCM has proposed using the CTF as part of MAPS to destroy appropriate NSCWM items found at APG.

Chemical Agent Munitions Disposal System

CAMDS, located at Deseret Chemical Depot near Tooele, Utah, is the Army's R&D facility for building and testing prototype chemical demilitarization hardware and processes. The demilitarization machines used in the stockpile CDFs and prototypes for the incinerators were fabricated and tested at CAMDS. It has been used by PMNSCM to develop, assemble, and test the RRS used for the disposal of CAIS. It has also been used to test systems for the biological degradation of chemical agents and is currently the Army's facility for the disposal of chemical materiel containing the arsenical agent lewisite. The lewisite that was stored in stockpile ton containers at Deseret Chemical Depot (DCD) was destroyed at CAMDS. Non-stockpile items containing lewisite (mortar rounds, projectiles, and a chemical sample bottle) stored at DCD are also intended for destruction at CAMDS.

The CAMDS physical facility consists of several buildings, incinerators, and engineering offices. As such, it is a valuable facility that can undertake specialized projects, destroy relatively small quantities of chemical agents, and develop and test equipment used for chemical munitions disposal.

Commercial Treatment, Storage, and Disposal Facilities

The fourth type of facility—commercial treatment, storage, and disposal facilities (TSDFs)—differs from stockpile and non-stockpile facilities in that commercial TSDFs cannot be used to treat CWM. However, they can accept the secondary waste generated by mobile systems and some individual treatment technologies, assuming that the secondary waste no longer contains agent, except perhaps at de minimis levels. A permit modification for treatment of these wastes may be required, however.

The Army has demonstrated two mobile systems for destruction of agent and energetics in NSCWM. These systems generate secondary wastes that can be disposed by other mobile systems or in treatment facilities that may be commercial or government owned. One is the RRS, designed to treat CAIS, which contain sulfur mustard, nitrogen mustard, and lewisite, and no energetics. The other is the EDS, which is designed to treat NSCWM with or without explosive components. Both are discussed later in this chapter.

The RRS was designed to neutralize the vast bulk of the agent, and the neutralent has been demonstrated to contain less than 1 ppm of agent.[3] The EDS has destroyed up to one-pound equivalent of explosives and reduced the concentration of agent (phosgene, mustard, and sarin) below detection

---

[3]John Gieseking, Office of PMNSCM, personal communication to R. Peter Stickles on January 15, 2002.

levels.[4] Both the RRS and the EDS use a liquid chemical formulation to neutralize (hydrolyze) the agent. In the EDS, the agent and energetics are converted into products that require further treatment before the mixture can be discharged (e.g., to a sewage treatment plant). In subsequent discussion, the committee refers to the liquid mixture produced in the RRS or the EDS as "neutralent."

Agent concentrations in neutralents from the RRS and EDS are so low that the neutralent should not be classified as chemical warfare materiel. Agent concentrations in rinsates and cleaning solutions from the EDS are even lower (see Table 2-2).

Table 2-2 demonstrates that the concentration of chemical constituents in the neutralent from the EDS are, in most cases, below federal land disposal restriction (LDR) treatment standards. While neutralents may contain very low levels of chemical agents, they present a risk similar to commercial hazardous waste and may be safely managed in commercial TSDFs. The destruction and removal efficiency (DRE) achieved by neutralization (99.9999 percent or more) exceeds or is comparable to the DRE achieved by the best commercial technologies. Thus, the committee believes that given the range of physical and chemical properties of the chemicals in neutralents and rinsates, these wastes could be safely managed in commercial TSDFs. State and/or federal regulators would need to agree, however, with the TSDF owner/operator that a particular neutralent or rinsate is appropriate for disposal at a given TSDF.

The use of commercial hazardous waste TSDFs for this purpose could reduce costs to the Army (and the taxpayer) with little or no adverse effect on human health or the environment. Informed consideration of the possibility of commercial treatment is consistent with the guidelines in Office of Management and Budget Circular A-76, which establishes the principle that the government should not perform functions that the private sector could perform unless there is a compelling reason to do so.

The committee also believes that commercially available hazardous waste incinerators should be suitable for final treatment of neutralents, although test burns may be necessary. Some neutralents are high in sodium, which tends to shorten the life of the refractory brick used to line incinerators, but wastes of similar composition have been treated satisfactorily. Commercial hazardous waste facilities are available that offer other technologies that might be better for aqueous wastes. These technologies include biological treatment, supercritical fluid extraction (not to be confused with supercritical water oxidation, discussed later in this chapter) followed by incineration of the smaller volume of extracted organics, and chemically based proprietary processes.

If rinsates and cleaning solutions are sufficiently dilute, they might be suitable for discharge directly to a publicly owned sewage treatment works (POTW) or a federally owned treatment works (FOTW). Rinsates and cleaning solutions sampled to date are generally dilute—more than 98.9 percent water. Except for the presence of chloroform from the sealant used in the EDS (NRC, 2001b, Table D-1), the levels of benzene, toluene, mercury, and arsenic in rinsate and cleaning solution are below RCRA LDR treatment levels and the levels allowed to be discharged to a POTW or FOTW (NRC, 2001b). Thus, based on a site-specific evaluation of the residual concentration of chemicals in the rinsate and cleaning solution waste streams, a POTW or an FOTW might accept these wastes for treatment. Of course, this determination should be made on a case-by-case basis and depends on the nature of the treatment system and the constituents and concentrations in the waste stream.

Nine hazardous waste incinerators that are operating commercially in the United States might be available, two each in Texas and Ohio, and one each in Arkansas, Illinois, Kentucky, Nebraska, and Utah. The largest commercial hazardous aqueous waste treatment facility in the United States is managed by DuPont in Deepwater, New Jersey. It provides a combination of physical, chemical, and biological treatment. Clean Harbors, in Baltimore, uses supercritical fluid extraction to treat aqueous wastes. Perma-Fix, with facilities in the Southeast and Midwest, uses proprietary aqueous treatment processes tailored to specific waste streams.

Status

Neutralents and rinsates from the RRS were destroyed in the Onyx Environmental Services commercial hazardous waste incinerator at Port Arthur, Texas. Other wastes from RRS operations were destroyed at Safety-Kleen's commercial hazardous waste incinerator in Aragonite, Utah, and

---

[4]The detection limits of <1 ppm (1 µg/ml ~ 1 ppm for dilute aqueous solutions) for phosgene and <0.2 ppm for mustard were lower than the treatment goal of <50 ppm for these two agents (U.S. Army, 2000b). Likewise, the detection limit of <0.1 ppm for sarin was lower than the treatment goal of <1 ppm (U.S. Army, 2001c). The treatment goals established by the Army for concentrations of residual agent in NSCM neutralents are not risk-based but rather appear to have been set in relation to detection limits. Nevertheless, toxicity studies conducted by the Army on RRS neutralents to date suggest that at residual mustard and lewisite concentrations of <50 ppm, the acute toxicity of the RRS neutralents is no greater than that of the virgin oxidant/solvent system. In Chapter 4, the committee indicates its belief that the approach exemplified by the Army's proposed Utah Chemical Agent Rule (UCAR) provides a good starting point for continued discussions between the Army, state regulators, and the public, on selecting appropriate standards that would be protective of human health and the environment.

TABLE 2-2 Composition of Liquid Waste Streams from the EDS Treatment of Sarin (GB) Bomblets at RMA

Waste Component from EDS Treatment of Sarin Bomblets at RMA	Neutralent[a,b]	Water Rinse[b]	Cleaning Solution[c]	POTW Feed Limit for Organic Chemical Industry	LDR Treatment Standards Goal of Treatment Prior to Disposal in Landfill
Monoethanolamine (MEA) (%)	43.8-48.3	0-4.7	0.8-1.1	None	None
Water (%)	51.7-56.2	95.3-100	98.9-99.2	NA	NA
IMPA (isopropyl methylphosphonic acid) (ppm)	3,400-5,000	24-78			NA
DIMP (diisopropyl methylphosphonate) (μg/L)	18,000-27,400	291-480	ND		NA
Explosives in liquids (μg/L)	<1,000	<1,000	<1,000		NA
Benzene (μg/L)	1,330-2,850	28.6-40.7	<100	137	140
Chloroform (μg/L)	ND-21.6	ND-4380	8,360-10,500	325	46
Dichloromethane (μg/L)	ND-97.1	ND-71	377-968		NA
Toluene (μg/L)	369-810	ND-23.7	<2	74	80
Mercury (μg/L)	0.1-1	0.1-2.65	17.9-25		150
Aluminum (μg/L)	8,720 to 11,100	876 to 11,800	<876		
Arsenic (μg/L)	<200	<20	<20		1,400
Cadmium (μg/L)	6.81-10	<0.68	2.2-46		690
Chromium (μg/L)	445-770	11.5-485	1,070-1,870		2,770
Copper (μg/L)	9,030-18,200	486-5,470	3,850-6,200		
Lead (μg/L)	63-237	3.82-603	128-168	690	690
Zinc (μg/L)	23,100-38,300	72.5-308	4,920-5,680	2,610	NA
pH	12	10.4-11.5	6.5-7.8		

NOTE: NA, not applicable; ND, none detected. The expected source and collection regime for these wastes are presented in Table 2-1. The term "treatment" is used to describe steps involving addition of reagent or water to the EDS and oscillating for some time period prior to opening the chamber. Note that water treatment and rinse water wastes can be combined. To date, the Army has chosen to segregate the three categories of wastes so as not to foreclose on the options for treating the waste streams that are primarily water.

[a]Neutralent consisting of the initial treatment of agent with active reagent (e.g., MEA) and any subsequent chamber washes with chemical reagent (if used).

[b]Rinsate consisting of additional agent treatment with water and chamber washes with water after opening the EDS.

[c]Cleaning solution consisting of washes (water and detergent) made between processing of each munition and final washes (e.g., water and acetic acid), made after completing a munitions campaign.

SOURCE: Lucille Forrest, Office of the PMNSCM, "Interpretation of Waste Results from EDS GB Bomblet Destruction, Rocky Mountain Arsenal," communication to the committee, February 2001.

Deer Park, Texas (U.S. Army, 2001d). Spent decontamination fluids generated at the Army's Dugway Proving Ground (DPG), Utah, have also been incinerated at the Aragonite facility. Chemical constituents in decontamination fluids are similar in general content to those in neutralents, but the concentrations of key contaminants are typically much lower.

Neutralents and rinsates from the EDS used at Rocky Mountain Arsenal have been destroyed at Safety-Kleen's commercial hazardous waste incinerator in Deer Park, Texas.

Technical Issues

Incineration technology is well developed. Destruction efficiency has been proven on wastes that are very similar in nature to the RRS and EDS neutralents and rinsates. Further, commercial incinerators have destroyed these specific wastes successfully. All commercial incinerators have elaborate air pollution control trains. Metal constituents of the wastes may remain with the slag or be captured by the air pollution control system. Both the slag and the fly ash are typically disposed of in a hazardous waste landfill.

The committee has concluded that CAIS items can be destroyed safely in hazardous waste incinerators, assuming the arsenic content does not present a site-specific air emission problem (NRC, 1999a). The physical and chemical properties of a chemical govern the ability of incinerators to destroy and remove the hazardous constituents. The hazardous constituents in CWM material share the chemical and physical characteristics of the hazardous constituents in commercial hazardous waste.

High-temperature commercial incineration is safe, robust (that is, applicable to a wide range of chemicals), and effective (that is, destroys or removes 99.99 percent or more of the chemicals treated) (EPA, 1999a, 1999b, 2000, 2001a, 2001b). It is also commercially available, heavily regulated, and widely utilized throughout the world, including for the destruction of hazardous wastes and chemical weapons.

However, as the committee has noted, a waste-specific (and in certain circumstances, a site-specific) determination of the safety and efficacy of incineration must be made before an incinerator can be permitted to treat non-stockpile chemical materials (NRC, 1999a). Such a risk assessment is required by EPA regulations and guidance on stack emission testing of combustion emission sources (EPA, 2001b).

Alternative technologies are available at several commercial hazardous waste facilities. To the committee's knowledge, none of these technologies has been used to treat neutralents or rinsates specifically. Though they have been used to treat similar types of wastes, a treatability study would probably be required prior to waste acceptance.

Regulatory Approval and Permitting Issues

A commercial hazardous waste facility can accept only wastes that are specifically included in its permit. Facilities try to include as broad a spectrum of waste as possible in their permits, because acceptance of a waste that is not included could require a permit modification.

Public Concerns

Some members of the public and several national environmental groups strongly oppose the incineration of both commercial hazardous waste and chemical weapons materiel. The nature and scope of opposition varies. For example, a public interest group with which some committee members met in Utah was very concerned about incineration at Tooele but was indifferent to the closure of a commercial incinerator in Clive, Utah. The congressional mandate to develop alternative treatment technologies stems from the strong opposition to incineration of the stockpile chemical weapons.

Ultimately, the Army and the state regulators must decide whether commercial facilities for the treatment or disposal of hazardous waste are the most appropriate approach after weighing all relevant factors, including technical feasibility, safety, legal and regulatory restrictions, willingness of the commercial facility to accept the material, timing, costs, and public concerns. Commercial rotary kiln incinerators have successfully destroyed secondary wastes from the destruction of non-stockpile materiel and are capable of handling a wide range of contaminated liquids, solids, and sludges. Other technologies offered by commercial hazardous waste facilities include deep well injection, biotreatment, and physicochemical treatment. It should be recognized that some members of the public may not believe that the risk posed by RRS and EDS neutralents and rinsates is minimal, but others may accept the trade-off between minimal risk and expeditious disposal of these liquid wastes. Thus, decisions about the acceptability of these technologies must be made on a case-by-case basis. The credibility of the Army and the likelihood of public acceptance (or acceptance by the majority of the public) are likely to be enhanced if the solicitation of public input is not viewed as a ploy to sell the public on a predetermined decision but rather as an open and transparent attempt to involve the public in government decision making. This approach results in a fair process; however, it does not and must not give veto power to persons who may still oppose the ultimate decision reached by the Army and state regulatory authorities.

Recommendation 2-4. The Army should continue to use commercial TSDFs or the stockpile facilities for the disposal of secondary wastes from the destruction of non-stockpile CWM when possible.

MOBILE TREATMENT SYSTEMS

The Army has developed two principal mobile systems for the treatment of non-stockpile CWM, the EDS and the RRS, and is developing a third but smaller system, the single CAIS accessing and neutralization system (SCANS). The EDS is designed to destroy explosively configured munitions, although it can also destroy chemical munitions without explosive components. The RRS is designed to dispose of CAIS at the locations where they are found. SCANS is being developed to treat individual CAIS vials or bottles. Another system, the Donovan blast chamber (DBC), was developed by a private corporation in Alabama and has been evaluated by the Army Corps of Engineers. The DBC, which was originally designed to treat conventional explosive munitions, has been modified and adapted to treat explosively configured CWM and potentially offers a higher rate of throughput than the EDS. Each of these systems is considered to be "mobile," although the term is a relative one. The

RRS, for example, requires three trailers or two C-141 aircraft to transport it. Both systems can be used as modular components of non-stockpile facilities or can be transported to the site of finds, as warranted.

Explosive Destruction System

The EDS is a trailer-mounted mobile system that is intended to destroy explosively configured chemical warfare munitions that are deemed to be unsafe to transport[5] or store routinely. It can also be used to destroy limited numbers of stable chemical munitions, with or without explosive components, when the quantity of these munitions does not require the use of other higher-capacity destruction systems. Technologies to destroy EDS secondary wastes were reviewed in an earlier report (NRC, 2001b). The EDS is described in Appendix E.

The EDS is being developed in two versions: a smaller Phase 1 (EDS-1) and a larger Phase 2 (EDS-2). The heart of the EDS system is an explosion containment vessel mounted on a flatbed trailer (see Figure E-1). The EDS-1 vessel's inside diameter is 20 inches (51 cm) and it is 36 inches (91 cm) long; the EDS-2 vessel is somewhat larger, with an inside diameter of 28 inches (71 cm) and a length of 56 inches (142 cm). The EDS-1 has an explosive capacity of 1 pound of TNT equivalent. The EDS-2 vessel will be capable of repeated use cycles at 3 pounds TNT equivalent and occasional uses at 5 pounds of TNT, should such a need arise. The frequency of allowable use above 3 pounds has yet to be determined. The vessel will be tested at more than 5 pounds of TNT equivalent for rating purposes.

Status

The EDS-1 (unit 1) was successfully used to destroy 10 M139 GB bomblets at Rocky Mountain Arsenal (RMA) in 2001. However, testing and improvement of the technology are still ongoing. The fabrication of two improved EDS-1 units (units 2 and 3) is scheduled for FY 2002. These will have a rotating chamber with an attached longitudinal baffle, or paddle, to improve the mixing of vessel contents during the neutralization reaction. Unit 1 of the EDS-2 design is under fabrication and will be tested for explosive capacity.

Deployment of the EDS is estimated to cost about the same as deployment of the RRS (~$2 million).[6]

Technical Issues

The EDS is intended primarily for the destruction of chemical munitions that contain fuzes or that are unsafe to transport or store long term. Since these munitions make up a small fraction of the NSCWM inventory at most sites (Table 2-3), the processing rate (currently one munition every 2 days) appears to be sufficient for the currently assigned mission and deployment plan. However, if the processing rate can be improved, the Army might have the flexibility to use the EDS to destroy a larger number of explosively configured and non-explosively-configured NSCWM. EDS-1 (unit 1) utilizes a rocking mechanism to achieve mixing of agent and reagent. Since the headspace of the EDS chamber is not thoroughly contacted, the processing time to obtain an acceptable headspace analysis is long. By employing rotation, the Army hopes to reduce the time to achieve effective neutralization (i.e., acceptable headspace analysis) of agent. The EDS-2's larger capacity (3 pounds of TNT equivalent vs. 1 pound TNT equivalent for the EDS-1) should allow the processing of multiple containers (e.g., Department of Transportation (DOT) bottles and vials of agent), thus increasing its nominal throughput capacity.

Owing to the length of the EDS-2 chamber, the longitudinal baffle cannot be fabricated by machining. The proposed solution is to weld the baffle to the inner surface of the vessel. According to Army sources,[7] postwelding heat treatment is not planned as part of the fabrication process. The heat-affected zone is a potential locus for accelerated chemical attack as a consequence of the residual stresses produced by welding.

Large safety factors have been built into the design of the EDS vessel and the procedures for its operation. The mechanical integrity of the vessel was evaluated by Sandia National Laboratories using a combination of small-scale failure analysis tests and computer simulations. This evaluation indicated that the EDS-1 containment vessel could withstand several thousand detonations with more than 1 pound of explosive, providing a significant margin of safety for a system with an intended system life of 500 detonations (Sandia National Laboratories, 2000).

Commercialization issues surrounding this technology were discussed in NRC (2001a). Definitive testing of the application of this technology to neutralent wastes was under

---

[5]The determination of whether or not a munition can be moved is made by Army Technical Escort personnel. Several factors are considered in making this decision, including (1) whether the munition is fuzed or unfuzed, (2) if fuzed, whether it is armed (i.e., if the munition was deployed as designed but failed to function properly), and (3) the severity of deterioration of the munition body and the physical state of the agent fill.

[6]John Gieseking, Office of PMNSCM, personal communication to R. Peter Stickles on January 15, 2002.

[7]Warren Taylor, Office of PMNSCM, personal communication to R. Peter Stickles on November 29, 2001.

TABLE 2-3 Number of Explosively Configured NSCWM and Total Recovered NSCWM, by Location

Location	Number of NSCWM	Number of Explosively Configured NSCWM
Aberdeen	125	8
Anniston	133	None
Blue Grass	4	None
Deseret	174	37
Newport	None	None
Pine Bluff	69,868	1,251[a]
Pueblo	12	None
Umatilla	5	None

[a]Only a small fraction of these munitions are fuzed such that they would need to be destroyed in an EDS.

SOURCE: Christopher Ross, PMNSCM, presentation to the committee on July 10, 2001.

way as this report was written, but the results were not available to the committee. This testing will determine the applicability of this technology to non-stockpile waste streams and identify the issues to be resolved in scaling up and commercializing the technology for these applications.

Regulatory Approval and Permitting Issues

Although the state of Colorado had issued a RCRA order pertaining to destruction of the GB bomblets at RMA, the Army conducted the EDS-1 cleanup operation at RMA as a Comprehensive Environmental Response, Compensation and Liability Act (CERCLA) emergency removal action (regulatory approval/permitting mechanisms are further discussed in Chapter 4 and Appendix F). This enabled the operation to proceed in a timely fashion, and the Army hopes that use of CERCLA emergency removal authority can be a model for future deployments of the EDS.

Public Concerns

The EDS appears to enjoy a good reputation among public interest groups. At RMA, it was the preferred option for destroying the GB bomblets of some members of the Core Group and of other public interest groups (see Chapter 5).

Finding 2-5. For the mission currently envisioned (i.e., disposing of limited numbers of explosively configured or nonexplosively configured NSCWM items), the EDS-1 and EDS-2 designs appear to be adequate, although units 2 and 3 of the EDS-1 and the larger-capacity EDS-2 need to be tested. Because it is applicable to a variety of non-stockpile materiel (see Chapter 3), the EDS is adaptable as one of the building blocks for facilities—for example, using a limited number of units operated in parallel.

Recommendation 2-5. The committee recommends that the Army continue to implement the planned improvements of the EDS that increase explosive capacity and reduce processing cycle time. The Army should consider the applicability of the EDS as modules in facilities.

Finding 2-6. The longitudinal baffle for EDS-2 will be attached by welding. Postwelding heat treatment is not currently anticipated.

Recommendation 2-6. The Army should engage appropriate technical resources to determine whether postwelding heat treatment should be considered to reduce the possibility of chemical attack in the heat-affected zone of the EDS-2.

Rapid Response System

The rapid response system (RRS) is a trailer-mounted chemical treatment system designed specifically to dispose of chemical agent identification sets (CAIS) at the locations where they are found. The operation of the RRS and technological options for destruction of secondary waste streams produced by the neutralization of CAIS were reviewed previously by this committee (NRC, 1999a, 2001a). The RRS can either be driven to or flown to locations where CAIS have been recovered. Transporting by air requires the use of two C-141 aircraft (one for the RRS operations and utility trailers and one for transporters), a supply trailer, and a mobile analytical support laboratory.

Description

The RRS contains a series of linked glove boxes equipped to remove CAIS vials and bottles from their packages, identify their contents, and neutralize those containing chemical agents (see Figure 2-2). CAIS containing sulfur mustard (H/HD), nitrogen mustard (HN-1 only), and lewisite (L) are chemically treated in the RRS (see Appendix E). CAIS containing industrial chemicals are segregated and repackaged for off-site commercial disposal.

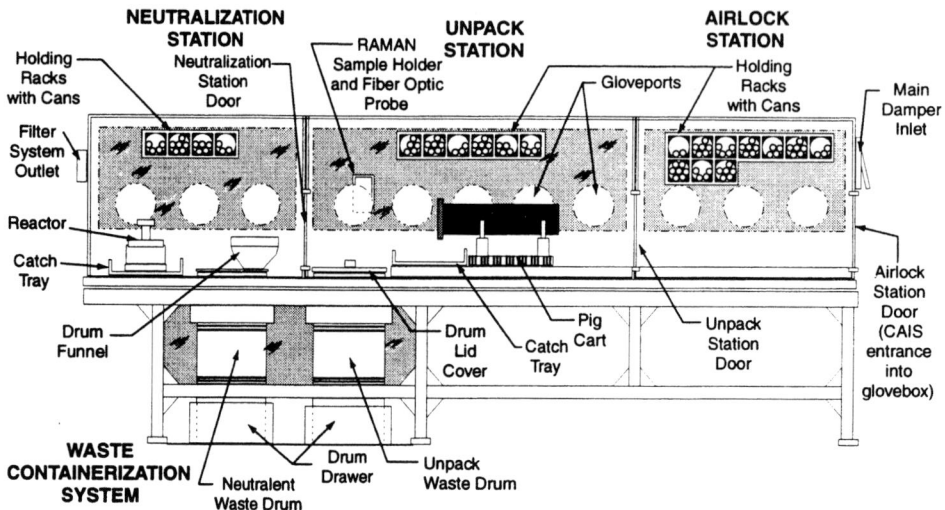

FIGURE 2-2 Glove-box system in the operations trailer of the RRS. SOURCE: U.S. Army (2001c).

Within the glove boxes, the glass containers are crushed in a reactor containing a chemical formulation that rapidly neutralizes the chemical agent. The contents of the reactor, including reagent, solvents, agent degradation products, and glass fragments, are transferred to sealed containers for disposal at a commercial TSDF. The RRS can treat one CAIS PIG[8] per day. More detailed information on the RRS appears in Appendix E.

The RRS is intended to be used at sites where many CAIS vials and/or PIGs containing CAIS sets are found. If only a few CAIS vials are found at a site, PMNSCM proposes to deploy the single CAIS accessing and neutralization system (SCANS) system once its development is complete (see below). The cost of transporting the RRS to a treatment site can be substantial and could affect the Army's disposal decisions. For example, the Army has estimated that deployment of the RRS to Fort Richardson, Alaska, a site with eight PIGs, would require 6 weeks and cost approximately $1.8 million (see Appendix E).

Status

A full-scale RRS prototype has been designed and constructed. The state of Utah approved a testing program to qualify the process, and 33 of the 60 sets of CAIS stored at Deseret Chemical Depot (DCD) were destroyed during this program. This operation was carried out successfully and is documented in detail (U.S. Army, 2001d). The goal of reducing agent concentration to less than 50 ppm was met, with most residue containers having agent concentrations of less than 1 ppm. The operations were then converted to a production mode, and the remaining CAIS at DCD—more than 1,200 items—were destroyed. Final reports on these operations were issued (U.S. Army, 2001d, 2001e).

Only one RRS is planned to be operational at a time. The RRS that the Army plans to build in 2006 will replace the existing unit. PMNSCM plans to make Pine Bluff Arsenal the home base of the RRS. Crews will be trained there and local CAIS will be destroyed when the RRS is not dispatched elsewhere.

Technical Issues

The committee is not aware of any major technical concerns with the RRS at this time. Following the successful CAIS treatment campaign at DCD, some modifications were made to the RRS by the contractor.[9]

Regulatory Approval and Permitting Issues

Currently, the RRS is permitted only for operations in Utah. Operating permits or alternative regulatory approvals

---

[8]PIG is a metal canister with packing material designed to protect CAIS during transport.

[9]The modifications were minor (e.g., increasing the capacity of the electric power supply system in response to a failure of the uninterruptible electric power supply system and moving nonessential equipment outside the operations trailer so as to reduce the internal heat load on the system) and do not affect the basic RRS operations.

(see Chapter 4) will have to be obtained from each state to which the RRS is deployed.

Public Concerns

During the operation of the RRS to destroy CAIS at DCD, some public interest groups raised concerns about the Army's plans to transport and destroy secondary waste streams from the RRS, which may contain trace amounts of agent as well as high concentrations of chloroform,[10] in a commercial incinerator.[11] Similar concerns may be raised at other sites to which the RRS is deployed.

Finding 2-7. The RRS is an expensive but adequate treatment system for CAIS PIGs and large numbers of loose CAIS vials and bottles. As other treatment options are also possible, this category appears to be well covered.

Single CAIS Accessing and Neutralization System

The Army is developing the SCANS (see Figure 2-3) to serve as a disposable neutralization reactor (i.e., a small pressure vessel) to treat individual CAIS vials or bottles. The SCANS could be deployed quickly to sites with only a limited number of CAIS vials or bottles, thus avoiding the time and expense associated with deployment of the RRS.

Description

The SCANS reactor is a small, disposable container used to access and treat CAIS vials or bottles containing chemical agents. Its process chemistry is similar to that of the RRS neutralization (see Appendix E). It is intended for use only where a limited number (estimated by PMNSCM to be 80 or fewer) of loose (uncontainerized) CAIS vials or glass bottles are recovered. Because SCANS does not have the glove box necessary to open a CAIS PIG safely, it could not be used for destruction of a CAIS PIG. SCANS also requires neither the elaborate system of trailers that supports the RRS nor its large operating crew and as such is much cheaper to deploy than the RRS.

As now planned, the SCANS would be used in conjunction with an analytical system such as the portable Raman spectrometer or the portable isotopic neutron spectrometer to identify the agents inside the vials or bottles. This step is necessary because the correct reagent must be selected to neutralize the agent in the bottle. Reagent would be added to the reactor, and a single CAIS item would be placed in the reactor. The reactor would be sealed, and a device such as a breaker rod would break the CAIS container. The agent would mix and react with the reagent to form a neutralent. The neutralent-containing reactor would be shipped to a permitted TSDF (which could include stockpile facilities) for disposal. If necessary, the reactor would be overpacked into a larger container meeting Department of Transportation requirements.

Status

At this writing, the SCANS reactor was undergoing technical feasibility testing; no decision regarding its fielding had yet been made. It is likely that SCANS would be used by the U.S. Army Technical Escort Unit or other designated agencies for accessing and treating CAIS vials during the first (or early) response to a discovery.

Technical Issues

The SCANS requires further engineering development, especially with regard to materials of construction; pressure, temperature, and reagent requirements; and selection and design of a breaker rod or similar system. Nevertheless, if it performs as anticipated, it should be an attractive and cost-effective system for treating a small number of individual CAIS vials or bottles. Further, the committee sees no reason the system could not be used at sites where larger numbers (e.g., dozens) of CAIS vials or bottles are found.

Regulatory Approval and Permitting Issues

Because of its intended use for small CAIS finds, the likely regulatory approval and permitting (RAP) mechanism for SCANS would be through RCRA emergency permits or CERCLA removal actions. Use of SCANS could also be approved through RCRA orders, especially at RCRA corrective action sites. If, as suggested by this committee, SCANS were used to treat a larger number of individual CAIS vials or bottles, these same RAP mechanisms could be employed. Even though the number of items would be somewhat greater than just a few vials or bottles, the nature and duration of the recovery action would still fit well within the scope of the less arduous RAP mechanisms. The committee could find no legal barrier to the use of these RAP procedures and expects that the state and federal regulators will also see the benefit of an expeditious, safe method for immediately neutralizing small quantities of agent and removing it for further treatment off-site, particularly when such finds are in residential areas or other areas to which the general public has access.

---

[10]Chloroform is a U044 hazardous waste. A concentration-based limit of 5.6 mg/kg has been set by the EPA for land disposal. The standard was established based on the performance of incineration, which is a best (demonstrated) available technology.

[11]Louise Dyson, Office of the Product Manager, Non-Stockpile Chemical Material, personal communication to Richard Ayen, July 10, 2001.

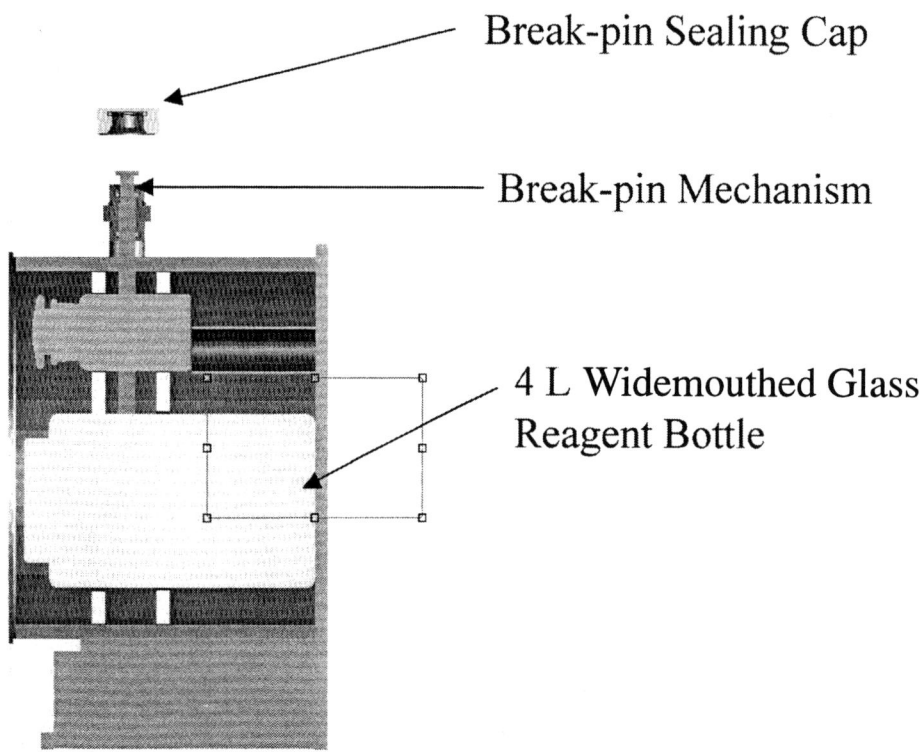

FIGURE 2-3 Schematic of one SCANS concept. SOURCE: Provided to the committee by Darryl Palmer, Office of the PMNSCM, February 27, 2002.

Compared with the option of overpacking and shipping recovered intact CAIS vials and bottles to a treatment facility (see Appendix G on transportation of NSCWM), SCANS offers the advantage of neutralizing the agent on-site and thereby reducing the concentration of chemical agents to de minimis levels and greatly reducing risk during transport.

Public Concerns

Because SCANS is expected to use the same basic neutralization and secondary waste treatment processes as the RRS, public concerns about the two are expected to be similar. However, SCANS provides a much faster response capability than does the RRS and a much smaller deployment footprint. These features are expected to be viewed as advantages by local public stakeholders at sites where individual CAIS vials or bottles are recovered.

Finding 2-8. The committee anticipates that SCANS will be a useful device, relatively low in cost compared with the RRS. If and when this potential is realized, several, and perhaps dozens, of SCANS units could be used to destroy the same number of CAIS vials or bottles safely, thus avoiding the time and expense of deploying the RRS.

Recommendation 2-8. The committee recommends that PMNSCM continue to develop and optimize SCANS to increase the number of CAIS vials and bottles that can be cost-effectively treated with multiple SCANS units. If the development program results in a system that can be cost-effectively used for a large number of vials and bottles, the system should be fielded as rapidly as possible. This approach would allow reserving the RRS for treating very large numbers of CAIS and PIGs containing CAIS, which the SCANS cannot process.

Donovan Blast Chamber

The Donovan blast chamber (DBC) was developed and is manufactured by DeMil International, Inc., of Huntsville, Alabama (DeMil International, 2001). The DBC was originally developed to replace conventional open detonation operations in a contained environment that prevents the release of blast fragments, heavy metals, and energetic by-products. It was later proposed that the DBC could be used to destroy CWM by detonation in its enclosed environment.

Description

The DBC consists of three main components: the blast chamber, an expansion chamber, and an emissions control unit, the latter consisting of a particle filter and a bank of activated carbon filters. The maximum explosive rating of the currently available T-10 mobile unit is 12 pounds of TNT equivalent. A larger mobile unit, which will have an explosive capacity of 50 pounds of TNT equivalent, is also under construction (DeMil International, 2001).

The blast chamber, in which the detonation occurs, is connected to a larger expansion chamber. A projectile wrapped in sheet explosive (either RDX or an aluminum-coated oxidizing composition) is mounted in the blast chamber. The floor of the chamber is covered with pea gravel, which absorbs some of the blast energy. The gravel is renewed periodically because it fractures during the explosions. Bags containing water are suspended near the projectile to help absorb blast energy and to produce steam, which reacts with agent vapors. After the blast chamber is loaded, its entry port is sealed and the exit from the expansion chamber is closed. After the sheet explosive is detonated, the chambers are kept sealed for about 2 minutes to maintain heat and pressure. The gases are then vented through the main duct to the baghouse and the carbon filters. Gases are monitored at several points in the DBC system and at the exit duct outlet. Particulates suspended in the vapors, such as soot, gravel dust, and metal oxides, are also monitored. Water vapor from the explosives and from the explosion-quenching water bags collects on the charcoal filters.[12]

The main waste materials from destroying chemical munitions are solids: soot, charcoal, gravel, inorganic dust, and metal fragments from the weapons. The only liquid waste from the DBC is spent hypochlorite solution from decontamination of the system prior to maintenance operations.[13]

The DBC in its currently fielded T-10 configuration appears to be able to treat complete CW munitions up to 105-mm, according to the manufacturer. The committee notes that the DBC could also be an appropriate treatment method for nonexplosive CWM, such as containers of agent or even quantities of CAIS vials or bottles.

Status

The use of the DBC to destroy chemical munitions was demonstrated in tests carried out in Belgium in May and June 2001. During those tests, live munitions containing the agents sulfur mustard, Clark, and phosgene were treated. The DBC system and operating procedure were modified to enhance worker safety and reduce potential emissions of residual chemical agent or agent decomposition products. Extensive monitoring was conducted to determine agent destruction efficiency and establish the quantity and nature of the decomposition products.[14]

During the Belgian tests, occasional breakthroughs of organic vapors from the filter system were detected. Further development will be necessary to ensure that no emissions occur. The solids, which are contaminated with chemical agent and explosives residues, were sent to a commercial hazardous waste incinerator for disposal.

After the detonation, the atmosphere in the blast chamber clears fairly rapidly, permitting reentry for maintenance and placement of the next round. During the tests in Belgium, 15 CW munitions were treated in the DBC in 3 hours, including 20-minute breaks after every 5 munitions (U.S. Army, 2001f). This amounts to an average treatment time of 12 minutes per munition, including the time for breaks. Analysis of the pea gravel and of wipe samples from the chamber walls consistently showed low agent concentrations during a test series. In the Belgian tests, all the residual solids, including the baghouse wastes, were sent to a commercial TSDF for thermal destruction or decontamination.

There appear to be mixed views of the usefulness of the DBC within the Army. The Corps of Engineers performed some limited testing of the DBC in the past using simulated agent (Zapata Engineering, 2000). The Corps appears to be in favor of testing the DBC further. PMNSCM has reservations about worker safety issues and overall permitability of the system (Stone & Webster, 2001a) and has no plans or budget for further testing.

Technical Issues

The DBC appears to be well suited for destroying a range of either chemical or conventional munitions. The destruction efficiency of just the explosion in the chamber appears to be about 99 percent.[15] The vast majority of the remaining 1 percent is either captured in the series of air pollution control devices or adheres to the pea gravel at the bottom of the DBC. Although only rarely was agent detected leaving the air pollution controls,[16] further development is necessary to ensure that no emissions of agent or recombinant products above the levels of concern occur.

Rather than washing the inside of the DBC with a neutralizing agent after every use, the Belgian DBC procedure de-

---

[12]The committee notes that water vapor competes with organic species for sites on the charcoal filters. Saturation of these sites with water vapor could reduce the effectiveness of the filters in removing organic species from the emission stream. One option for addressing this potential problem would be to add a condensation step to the emission control system before the filtration step.

[13]Herbert C. De Bisschop, personal communication to G.W. Parshall in an interview at the Belgian Royal Military Academy, July 25, 2001.

[14]Herbert C. De Bisschop, personal communication to G.W. Parshall in an interview at the Belgian Royal Military Academy, July 25, 2001.

[15]Herbert C. De Bisschop, personal communication to G.W. Parshall in an interview at the Belgian Royal Military Academy, July 25, 2001.

[16]Herbert C. De Bisschop, personal communication to G.W. Parshall in an interview at the Belgian Royal Military Academy, July 25, 2001.

stroys CWM for several days, then decontaminates the inside of the chamber and the pea gravel with sodium hypochlorite solution (a highly efficient neutralizing agent). The washed pea gravel is further decontaminated in an incinerator. Based on the fundamental chemistry of these materials, the hypochlorite should destroy the agent in the chamber, including nearly all of that associated with the pea gravel. As a result, it appears that a very high proportion of the chemical agent would be destroyed in the explosion, neutralized by the hypochlorite, or adsorbed to the carbon and other filters in the air pollution control system. All residual solids, including carbon and other filtered waste, would then be destroyed or decontaminated by thermal treatment. Disposal of carbon filters continues to require further review for all technologies. It may be necessary to neutralize the gravel and dust before shipment off-site. Thus, the DRE for the complete DBC system may be comparable to that for other treatment systems used for chemical agent destruction. However, it may be necessary to conduct further testing to establish a quantitative estimate of the DRE. The committee has not yet received the joint Belgian-American analytical report from the May-June 2001 tests.

Because there is no time-consuming neutralization step, the DBC's throughput appears to be much higher than that of the EDS, which treats only one munition every other day. The DBC also has the advantage of generating little liquid waste that requires subsequent processing, in contrast with the substantial neutralent and rinsate effluents produced with use of the EDS.

Regulatory Approval and Permitting Issues

The DBC has not been permitted for use in destroying CWM in the United States, although it has been used successfully in Europe. Additional testing of the DBC will be required if the system is to be permitted in the United States for treatment of CWM. While the system's DRE, at 99 percent, is comparatively low, the DRE of the entire system, including hypochlorite decontamination and further treatment of solids (e.g., via incineration), would need to be considered.

Additionally, the committee believes that more data are needed on the likelihood that undestroyed agent or other hazardous constituents could be released when the DBC door is opened between uses and prior to the periodic washing. The chamber is kept under negative pressure, which means that air flows from outside the chamber into it. However, the reliability and efficiency of these systems need to be documented and provided to regulators if approval is sought.

Public Concerns

Public concerns about the DBC are not known at this time. In the one case in which use of the DBC was proposed for emergency removal of non-stockpile CWM (the GB bomblets recovered at Rocky Mountain Arsenal (RMA)), public interest groups expressed a preference for the EDS instead.

Finding 2-9. In operational tests conducted in Belgium, the Donovan blast chamber (DBC) processed at least 15 chemical munitions per 8-hour day (typically 75-mm projectiles). For operation in the United States, further controls may be required and may reduce throughput. This chamber would be useful at sites containing large numbers of CWM. However, additional information on destruction and removal efficiencies, agent containment, and worker safety is needed. Secondary waste treatment must also be taken into account. If the DBC is to be considered further, more testing will be required to establish that it can meet U.S. regulatory requirements.

Recommendation 2-9. The non-stockpile program should continue to monitor the Belgian tests of the DBC. If the results are encouraging and it appears that the DBC can be permitted in the United States, it should be considered for use at sites where prompt disposal of large numbers of munitions is required.

INDIVIDUAL TREATMENT TECHNOLOGIES

The treatment facilities and systems discussed in the previous section involve a combination of technologies, including for the preparation of a munition for processing, agent accessing, agent destruction, and treatment of secondary waste materials. The Army has available a variety of individual treatment technologies that can be utilized on their own or integrated into the systems and facilities, as discussed above, to accomplish specific tasks.

Some of the individual treatment technologies are mature and have been used for years for NSCWM disposal; others are mature commercial technologies that have not been fully tested for application to NSCWM. Finally, some are still at the developmental stage. PMNSCM is continuing to implement its technology test program to investigate the suitability of several of these technologies for destruction of NSCWM or secondary waste streams. The technologies being evaluated are listed in Table 2-4 and are discussed further below.

Plasma Arc

Plasma arc is a very high temperature process that could be used to destroy neat agent or secondary waste streams resulting from agent neutralization. It is also suitable for destroying metal parts, dunnage, and energetics.

Description

Plasma arc technology utilizes the electrical discharge of a gas to produce a field of intense radiant energy and high-

TABLE 2-4 NSCMP Technology Test Program

Technology	Vendor, Test Site	Feed Streams
PLASMOX	Burns and Roe Enterprises, MGC Plasma in Muttenz, Switzerland	H neutralent simulant, GB neutralent simulant
Gas-phase chemical reduction	Eco Logic International, Inc., Edgewood, Maryland	GB neutralent, H neutralent simulant, RRS neutralent, DF simulant, vials of $CHCl_3$
Supercritical water oxidation (continuous)	Both vendor and test site to be determined	Binary chemicals, rinsates, neutralents
Supercritical water oxidation (batch)	Sandia National Laboratories, Livermore, California	H neutralent simulant, GB neutralent simulant, vials of $CHCl_3$
Persulfate oxidation	Southwest Research Institute, San Antonio, Texas	HD neutralent simulant, GB neutralent simulant, DF
Electrochemical oxidation	CerOx Corporation, University of Nevada at Reno	H neutralent simulant, GB neutralent simulant, DF simulant
Ultraviolet oxidation	Purifics Inc., Toronto, Canada	Rinsate simulant
Wet air oxidation	Zimpro Products, Rothschild, Wis.	Neutralent simulant, binary DF, and QL simulant

SOURCE: Christopher Ross, PMNSCM, presentation to the committee on July 10, 2001.

temperature ions and electrons that cause target chemical compounds to dissociate in a containment chamber. Plasma arc generates large volumes of high-temperature vapor that require high-quality treatment.

There are many variations of the plasma arc process, involving use of different plasma gases and reactor designs that provide either an oxidizing or a reducing environment. One system developed by MGC Plasma AG in Switzerland (the PLASMOX process) has achieved destruction efficiencies greater than seven nines (99.99999 percent) when processing adamsite, Clark I and II, phosgene, lewisite, yperite, and a mixture of yperite and lewisite. PLASMOX employs closely coupled, staged reaction zones (characterized as controlled pyrolysis) to completely destroy organic compounds. The Army has also investigated the PLASMOX process for destruction of neutralent waste streams as part of its technology test program.

Status

MGC/PLASMOX developed a portable unit, Model RIF 2, that was put into operation in 1994, and it has since built additional units. The RIF 2 is skid-mounted and designed to be moved by four standard tractor-trailers. The unit has been used in Europe and is permitted under both Swiss and German environmental laws and regulations. It was used successfully to destroy chemical agents for the Swiss Army at its chemical materiel laboratory in Spiez, Switzerland. The PLASMOX tests run by the Germans and Swiss indicate that the system will destroy chemical warfare agent safely and rapidly (Burns and Roe, 2001).

There has been no recorded destruction of NSCWM by plasma arc technology in the United States; however, as part of the technology test program (see Table 2-4), PMNSCM hired Stone & Webster to conduct tests of the Burns and Roe PLASMOX plasma arc process on simulated H and GB neutralents with MEA. MGC conducted these tests from January 8 through January 19, 2001, under a subcontract to Burns and Roe Enterprises at the MGC/PLASMOX facility in Switzerland. The system layout is shown in Figure 2-4.

PMNSCM has proposed that plasma arc technology be used primarily for the destruction of neutralent waste streams, although it may be a candidate for the direct destruction of the binary CWM components DF and QL stored at Pine Bluff Arsenal. Based on the MGC/PLASMOX tests, the throughput rate for neutralent processing is approximately 13 liters per hour, with a 50 percent availability.

Technical Issues

Stone & Webster recommended that the PLASMOX system receive further testing on typical NSCMP liquid and solid waste streams, with particular attention paid to the deposition of solid materials in the system. Its report concluded that further improvements would have to be made to ensure that the system would comply with all EPA and state requirements (Stone & Webster, 2001b).

The Army has identified approximately a dozen vendors of plasma arc technology in the United States, although none is currently permitted to treat hazardous waste or NSCWM.

Regulatory Approval and Permitting Issues

The Army's test results for the PLASMOX technology raise a number of regulatory issues that must be resolved before this system could be permitted in the United States. These include improvements to the gas scrubber system, more complete knowledge of the fate of key components of the NSCWM (e.g., phosphorus), and better characterization of the solid, liquid, and gaseous waste streams. The processes used by U.S. plasma arc vendors also differ in significant ways from the PLASMOX process tested by the Army.

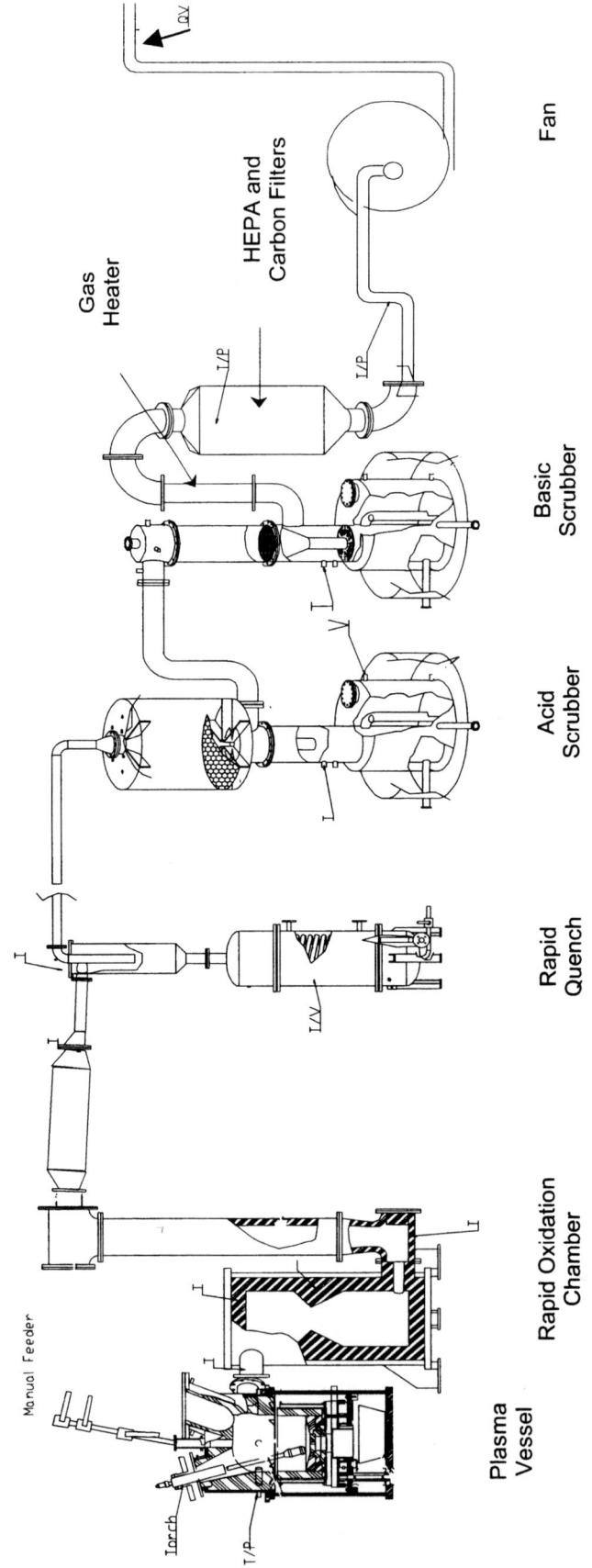

FIGURE 2-4 PLASMOX system layout. SOURCE: Adapted from Stone & Webster (2001b).

Public Concerns

A key public concern about plasma arc processes for the destruction of NSCWM in the United States is whether plasma arc offers a true alternative to incineration. Depending on the type of plasma gas used and the configuration of secondary oxidation zones, quench, and scrubber processes, plasma arc systems may produce gas volumes and reaction products that are quite similar to or quite different from those associated with incinerators. Plasma arc processes that use oxygen as the plasma gas and/or have postcombustion chambers may be practically indistinguishable from incineration. On the other hand, a case can be made that plasma arc processes that do not use oxygen as the primary plasma gas differ from incineration, although even in these systems, oxidation generally takes place at a subsequent stage of the process. However, when the plasma arc system is optimally designed and controlled, dioxins, furans, and other hazardous pollutants are likely to be below regulatory limits.

One indication of public attitudes toward plasma arc is that after careful consideration, the Assembled Chemical Weapons Assessment (ACWA) program Dialogue Group (NRC, 1999b) accepted plasma arc as a valid alternative to incineration. However, a spokesperson for the Non-Stockpile Chemical Weapons Citizens Coalition characterized plasma arc as a synonym for incineration and expressed concern that NSCMP was prematurely embracing the technology. As with incineration, the degree of public concern about plasma arc may vary with specific implementation and specific location.

Finding 2-10. At least some plasma arc systems appear to be robust technology capable of highly efficient destruction of a variety of non-stockpile agents and secondary waste streams in a safe, environmentally acceptable manner. Systems that employ closely coupled reducing and oxidizing zones (controlled pyrolysis) have produced good results in demonstration and actual destruction projects in Europe. Plasma systems may require permitting as incinerators and may raise related public concerns.

Recommendation 2-10. Additional testing of plasma arc technology should be done to ensure that proposed plasma arc systems are capable of meeting the requirements of the Environmental Protection Agency (EPA) and state requirements.

Chemical Oxidation

The use of chemical oxidation to treat liquid secondary waste streams from the RRS and EDS has been discussed extensively in previous reports by this committee (NRC, 2001a, 2001c).

Description

The use of hydrogen peroxide or Fenton's reagent, potassium permanganate, Oxone,[17] peroxydisulfate, peroxyborate, peroxycarbonate, peroxyphthalate, and UV-activated hydrogen peroxide or ozone oxidation is a promising approach for the treatment of liquid waste streams because of their demonstrated technical effectiveness for similar waste streams, good pollution prevention qualities, and low cost. The reactions are carried out at 80°C to 100°C at atmospheric pressure in aqueous solutions. Under appropriate conditions, the organic constituents of the neutralents and rinsates can be mineralized. In other cases, they are converted to less active compounds.

Oxidation without UV activation is preferred. The problems associated with UV activation are discussed in *Disposal of Neutralent Wastes* (NRC, 2001c). These include the need for special equipment, reduced effectiveness for opaque solutions, and fouling of the optical surfaces. For chemical oxidation not catalyzed by UV light, conventional chemical process equipment and procedures are used.

Status

Chemical oxidation is a mature process widely used in the chemical industry. Until recently, however, there was no direct experience with the chemical oxidation of the liquid products resulting from treatment of the agent in non-stockpile weapons. UV-activated hydrogen peroxide oxidation was tested in 2001 by the PMNSCM on a 2 percent MEA rinsate simulant using equipment and facilities supplied by Purifics Environmental Technologies.[18] Very limited information is available. With the application of 1,550 kWh/m^3 of power, the total organic carbon of 286 liters of rinsate was reduced from 10,000 mg/L to about 9 mg/L over a period of 92 hours.

The committee was not aware of any reported results of oxidation of neutralent or rinsates without UV activation. The Army is testing persulfate oxidation of neutralents at the Southwest Research Institute (Table 2-4), but no results were available at the time this report was written. However, the successful oxidation or mineralization of closely related materials, including mustard and nerve agents and their hydrolysates, has been documented. Workers at the U.S. Army Edgewood Research, Development, and Engineering Center (Yang, 1995) have conducted laboratory-scale studies on the reaction of VX, GB, GD (soman), and mustard with hydro-

---

[17]Oxone, a registered trademark of DuPont Specialty Chemicals, is a triple salt (2KHSO$_5$·KHSO$_4$·K$_2$SO$_4$).

[18]Edward Doyle and Joseph Cardito, Technology Test Program for Treatment of NSCMP Feeds, presentation to the committee on September 25, 2001.

gen peroxide, Oxone, hypochlorite, and peroxydisulfate; they had very favorable results, especially with peroxydisulfate.

The committee notes that unlike most other technologies being considered, chemical oxidation without UV activation is not promoted by any technology vendors. The technology is not proprietary, and any organization can design and build the equipment, purchase the reagents, and carry out the operations. Accordingly, it would be up to the Army to take the initiative in the exploration and development of this technology.

Technical Issues

The greatest potential disadvantage of chemical oxidation is that it may not fully mineralize all of the compounds in the neutralents or may not mineralize them rapidly enough to be practical. Many organics, particularly simple aliphatics and halogenated alkanes, are somewhat recalcitrant to simple chemical oxidation. Long reaction times and large amounts of oxidant may be required to achieve a satisfactory result. Only testing can resolve these issues.

The commercialization issues surrounding this technology were discussed in NRC (2001c). Definitive testing of the application of this technology to neutralent wastes was under way as this report was written, but the results were not available to the committee. The testing will determine the applicability of this technology to non-stockpile waste streams and identify the issues to be resolved in scaling it up and commercializing it for these applications.

Regulatory Approval and Permitting Issues

Provided that chemical oxidation can be demonstrated to be effective in destroying NSCWM liquid secondary waste streams, no particular problems are anticipated in obtaining the necessary regulatory approvals.

Public Concerns

Public reaction to the use of chemical oxidation of NSCWM secondary waste streams as an alternative to incineration is expected to be very favorable. Emissions from the process are minimal, and the formation of chlorodibenzodioxins and chlorodibenzofurans is expected to be unlikely at low temperatures (this should be verified, however).

Wet Air Oxidation

The use of wet air oxidation (WAO) to treat liquid secondary waste streams of the RRS and EDS was discussed extensively in previous reports by this committee (NRC, 2001a, 2001b).

Description

WAO is a hydrothermal process for the oxidative destruction of organic wastes that is carried out at 150°C to 315°C and 150 to 3,000 pounds per square inch, absolute (psia). The oxidizing agent is dissolved oxygen. WAO is a mature commercial technology that is used widely in the United States to treat wastewater and various hazardous waste streams.

WAO can treat any pumpable fluids provided that the chemical oxygen demand is less than 120,000 mg/L. In the case of NSCWM neutralents, for example, it is estimated that this would require a seven- to ninefold dilution with water.[19] Such dilution could be achieved in whole or in part by combining the primary neutralent with the dilute aqueous rinsates.

Status

WAO is used routinely in commercial applications to treat sewage sludge containing 10 to 15 percent solids. WAO does not fully mineralize organics but instead reduces them to short-chain molecules such as acetic acid. Thus, effluents may have to be treated further by biotreatment, possibly at a POTW. Prior to biotreatment, toxic heavy metals would have to be removed. Arsenic, if present, would be converted to arsenate ion, a form that is more readily stabilized. NSCMP has plans to test WAO for treatment of secondary waste streams (see Table 2-4).

Technical Issues

As with chemical oxidation, the principal uncertainty surrounding the application of WAO to NSCWM liquid waste streams is whether it is capable of converting the various compounds of concern into materials that are acceptable to a POTW. Planned testing of this technology by the Army should help to resolve this question.

The commercialization issues surrounding this technology were discussed in (NRC, 2001a). Definitive testing of the application of this technology to neutralent wastes was under way as this report was written, but the results were not available to the committee. The testing will determine the applicability of this technology to non-stockpile waste streams and identify the issues to be resolved in scaling it up and commercializing it for these applications.

Regulatory Approval and Permitting Issues

If WAO can be demonstrated to be effective for NSCWM liquid waste streams, no particular problems are anticipated in obtaining the necessary regulatory approvals. However,

---

[19]William Copa, U.S. Filter Zimpro, personal communication to Joan Berkowitz on February 15, 2000.

because RCRA prohibits dilution as a means of achieving treatment standards, the Army may need to demonstrate that the dilution that is inherent to this process is necessary to achieve the appropriate conditions under which the technology normally operates.

Public Concerns

No particular public concerns regarding the use of WAO to treat NSCWM secondary waste streams are anticipated, based on its current widespread use as a wastewater treatment technology.

Batch Supercritical Water Oxidation

The use of batch supercritical water oxidation (batch SCWO) to treat liquid secondary waste streams of the EDS was reviewed in a previous report by this committee (NRC, 2001b).

Description

A batch SCWO unit is conceptually similar to a pressure cooker. Material to be destroyed is mixed with an oxidizer (such as hydrogen peroxide) and introduced into a pressure vessel, which is heated to a reaction temperature above the critical point of water (374°C, 3,204 psia) and then cooled. Organic materials are mineralized to produce carbon dioxide, nitrogen, and aqueous salts.

In principle, batch SCWO could be used to treat either neat agents or more dilute secondary liquid waste streams from chemical neutralization processes. Although both applications were initially considered, the Army is no longer considering the latter, because the volumetric throughput of the batch SCWO process is low compared with the volume of liquid wastes that would be generated by treatment systems such as the EDS. Batch SCWO is still being considered for direct treatment of CAIS vials and bottles and, in the longer term, to replace the neutralization of agent released from chemical munitions following explosive accessing inside the EDS containment vessel. The advantage of the latter process would be that all operations are carried out in one vessel and that it avoids the production of secondary liquid waste streams requiring further treatment.

Status

Four bench-scale batch SCWO reactors have been constructed at Sandia National Laboratories in Livermore, California, each having a volume of 325 ml. Both H and GB neutralent simulants have been processed in the units, achieving a destruction and removal efficiency (DRE)[20] of 99.988 percent.

---

[20]DRE is calculated as the percentage of agent destroyed or removed.

An experiment to test the feasibility of using batch SCWO to destroy CAIS vials was conducted, in which a half-sized simulated CAIS vial (a vial containing neat chloroform) was placed in the batch SCWO. The CAIS vial burst open when the temperature reached about 300°C, demonstrating that no additional device would be needed to access the contents of CAIS vials. A prototype of a combined EDS-batch SCWO does not yet exist. For a combined EDS-batch SCWO reactor, the same 20-inch inner diameter would allow use of the existing EDS-1 door, hinges, and clamps. The length of the vessel would be doubled, from 39 to 78 inches. Throughput rates would depend on the time required for SCWO operations, currently estimated to be 2 to 3 hours, including heating and cooling. The combined EDS-batch SCWO would be able to process up to a 4.2-inch mortar round.

Much more testing and scale-up work must be done. The combined EDS-batch SCWO is part of the long-term technology program for FY 2002-2009, with prototype testing scheduled between July 2005 and July 2006.

Technical Issues

Batch SCWO is part of NSCMP's long-term technology test program (see Table 2-4). As such, several years of development and scale-up are required before an operational unit is available for testing. Decisions concerning subsequent deployment of this unit would depend on the needs at the time that it is available (mid-2006, at the earliest).

Preliminary results on the direct destruction of simulated CAIS vials in a batch SCWO reactor appear promising; however, it remains unclear how widely applicable this approach is to the range of vials and bottles (and the wide range of vial contents) that make up CAIS sets (NRC, 1999a). Direct treatment of CAIS in a batch SCWO has the advantage that no secondary waste streams that require further treatment would be generated; however, the cost-effectiveness of this approach, especially relative to the SCANS plus secondary waste treatment, is unclear.

For the combined EDS-batch SCWO, a variety of technical issues must be addressed: the choice of materials of construction, the method used to introduce oxidant into the vessel, the durability of seals, the stability of SCWO reactions in a large-diameter vessel, the methods used to heat the vessel, the possibility of scaling and corrosion under batch SCWO conditions (salts are proposed to be captured in a pan placed in the vessel, but this has not yet been demonstrated), the method used to fabricate the vessel (e.g., single forging vs. welded sections), the impact of repeated explosions followed by thermal and pressure cycles on the integrity of the EDS vessel and SCWO reactor (e.g., crack propagation), the most appropriate method of cooldown and depressurization following munition destruction, and the disposition of process residuals.

In addition to the technical issues noted above, the Army would have to demonstrate that a combined EDS-batch

SCWO unit is more cost-effective than alternatives and make a case for the "market" for this unit (i.e., show that the number of non-stockpile items that it will dispose of justifies the costs of technology development).

Given the late date of availability of batch SCWO and the existence of alternatives (e.g., EDS followed by incineration of neutralents), the committee questions the cost-effectiveness of this technology and the role that it will have in the destruction of recovered or buried non-stockpile materiel.

Other Alternative Individual Treatment Technologies

In previous reports (NRC, 2001a, 2001b), the committee evaluated several other alternative technologies that have been considered by the Army for treatment of non-stockpile secondary liquid waste streams, including electrochemical oxidation, continuous mode SCWO, gas-phase chemical reduction, biotreatment, and solvated electron technology. These technologies were evaluated against a number of criteria, including technical effectiveness, safety, permit status, and pollution prevention attributes. All were found to have serious deficiencies in one area or another, especially in comparison with the technologies that the committee considered most promising: chemical oxidation, wet air oxidation, and plasma arc. Consequently, the committee recommended that no further investment be made in developing electrochemical oxidation, supercritical water oxidation (continuous mode), gas-phase chemical reduction, biotreatment, or solvated electron technology for the treatment of non-stockpile secondary liquid waste streams. During the preparation of this report, the committee reviewed its earlier analysis and found nothing that would change its conclusions.

Finding 2-11. The Army has considered three possible applications of batch SCWO:

1. as a treatment method for liquid secondary waste streams from the RRS or EDS
2. as a direct treatment method for CAIS vials or bottles, instead of the RRS or SCANS
3. as an alternative to chemical neutralization in the EDS explosion containment chamber

The low volumetric throughput of batch SCWO appears to make it inappropriate for option 1. The advantages of option 2 over existing treatment alternatives are unclear. While option 3 offers the possibility of eliminating the generation of neutralent, the committee sees significant technical issues associated with the thermal and pressure cycling of the EDS vessel for SCWO operation. This technical immaturity makes it unlikely that option 3 could contribute to the Army's NSCWM destruction mission by PMNSCM's completion goal of April 29, 2007.

Recommendation 2-11. The Army should continue R&D on batch SCWO only if it can demonstrate that the technology is more cost-effective than alternatives and that the number of non-stockpile items it will dispose of justifies the costs of technology development.

Finding 2-12a. The Army's plan to destroy highly organic neutralent waste streams by incineration is appropriate. Plasma arc systems are also adaptable to destruction of highly organic neutralents when incineration is not available or acceptable. Use of such high-temperature processes to destroy aqueous secondary wastes would be inefficient, although it may be expedient in some cases. If such aqueous liquids cannot be disposed via publicly or federally owned treatment works (POTW or FOTW), chemical oxidation or wet air oxidation may be attractive alternatives for this purpose.

Finding 2-12b. Several other alternative treatment technologies being investigated by PMNSCM appear to be less appropriate for the needs of the program. The current analysis points out the need for more information on the scope of applicability of UV-catalyzed oxidation and calls into question the need to continue development of batch SCWO processing. The potential applicability of UV oxidation is not well understood, and batch SCWO, when combined with explosive accessing in an EDS, is technologically immature.

Recommendation 2-12a. The PMNSCM should continue its research and development program on chemical oxidation and wet air oxidation of neutralents and rinsates.

Recommendation 2-12b. Consistent with the committee's earlier analyses (NRC, 2001a, 2001b) there should be no further funding for the development of biological treatments, electrochemical oxidation, gas-phase chemical reduction, solvated electron technology, and continuous SCWO technologies for the treatment of neutralents and rinsates. PMNSCM should monitor progress in technologies being developed under the Assembled Chemical Weapons Assessment (ACWA) program but should evaluate ACWA technologies for the treatment of non-stockpile neutralents and rinsates only if no additional investment is required.

Neutralization (Chemical Hydrolysis)

One potentially attractive approach to the destruction of the chemical warfare agents is chemical hydrolysis reaction with water to form products of reduced toxicity. In a chemical demilitarization context, the process is known as "neutralization," because it neutralizes the toxic properties of the chemical agent.

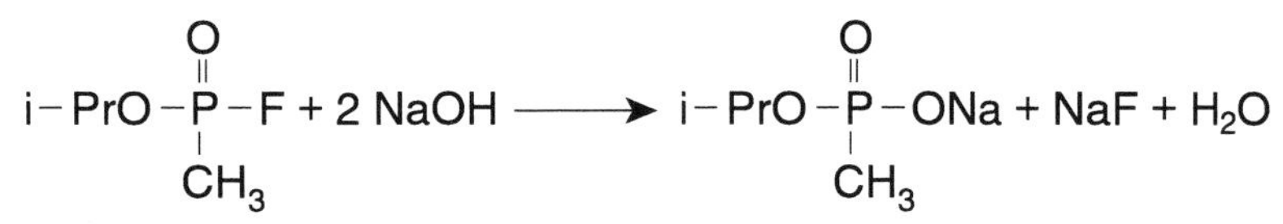

$$\text{i-PrO}-\underset{\underset{CH_3}{|}}{\overset{\overset{O}{\|}}{P}}-F + 2\,NaOH \longrightarrow \text{i-PrO}-\underset{\underset{CH_3}{|}}{\overset{\overset{O}{\|}}{P}}-ONa + NaF + H_2O$$

GB (sarin)   Isopropyl methylphosphonic acid (IMPA) salt

FIGURE 2-5 Hydrolysis of the nerve agent GB (sarin). SOURCE: Flamm (1987).

Description

Hydrolysis is usually carried out under mild conditions, typically atmospheric pressure and less than 100°C, in conventional chemical process equipment. Because some of the hydrolysis products are acids, a base such as lime or sodium hydroxide is often used to chemically neutralize the reaction mixture. The base may also accelerate the hydrolysis, permitting it to be carried out more rapidly and with more efficient use of process equipment. Neutralization is potentially applicable to the whole family of phosphate-based nerve agents, to blister agents, and to binary precursors of nerve agents. The hydrolysis of the nerve agent GB (sarin) has been studied extensively and is described by the equation in Figure 2-5. The main neutralization processes in the non-stockpile program utilize aqueous monoethanolamine (MEA) as the neutralizing reagent for HD, GB, and VX agents. One of the virtues of this approach is that the neutralents thus produced are single-phase liquids.[21]

After neutralization, the products are neutral or alkaline solutions containing inorganic salts and organic compounds of greatly reduced toxicity. The product solutions can be treated further to produce "mineralized" products suitable for disposal in the same way as normal chemical wastes. Some such treatment is required to destroy gross quantities of CWC Schedule 2 chemicals (e.g., the sodium salt of IMPA in the above reaction) so that they cannot be converted back to chemical agents.

Status

The U.S. Army's experience with neutralization of the nerve agent GB confirmed the potential virtues of this technology, as well as some practical problems associated with it. In an extensive field program carried out from 1973 to 1976 at the Rocky Mountain Arsenal, 4,188 tons of GB were hydrolyzed successfully (Flamm et al., 1987). The nerve agent was treated with a large excess of aqueous sodium hydroxide to produce a water solution of inorganic salts and organic compounds. The solutions were evaporated and the solid residues deposited in a hazardous waste landfill. With hindsight, it appears that working with a smaller ratio of alkali to GB would have substantially reduced the amount of solid wastes produced in the campaign. In addition to the U.S. Army experience with hydrolysis of GB, various neutralization processes have also been used to destroy multiton quantities of the agent in Great Britain, Canada, the countries of the former Soviet Union, and Iraq. Overall, chemical hydrolysis is attractive for the destruction of chemical agents and their precursors because it is simple and well proven, uses standard commercial process equipment, and operates under mild temperature and pressure.

Neutralization of Binary Components

Chemical neutralization is being considered for disposal of the PBA inventory of the chemicals DF (CH_3POF_2) and QL ($CH_3P(OEt)(OCH_2CH_2N\text{-}i\text{-}Pr_2)$), which are precursors of the nerve agents GB and VX, respectively. Both compounds are highly toxic although much less lethal than GB and VX.

Both DF and QL were destroyed by hydrolysis on a significant scale in a campaign at Aberdeen Proving Ground in 1997.[22] As seen in Figure 2-6, hydrolysis of DF with warm water yielded an aqueous solution of hydrogen fluoride and methylphosphonic acid (MPA) and its derivatives.

Like the caustic hydrolysis of GB, the reaction of DF with water proceeded exothermally. The DF neutralent was sent to a commercial TSDF for disposal.

Implementation of binary precursor neutralization does not appear in current NSCWM disposal budgets. A pilot test program for neutralization of DF and QL was scheduled to begin in late September 2001 (see Table 2-4) but has not started due to restrictions on the movement of these two

---

[21]Lucille Forrest and James Horton, Office of the PMNSCM, personal communication to G.W. Parshall on October 1, 2001.

[22]Christopher Ross, PMNSCM, presentation to the committee, November 8, 2001.

$$DF + H_2O \longrightarrow HF + CH_3P(O)(OH)_2 + CH_3PF(O)(OH)$$

Methylphosphonic acid (MPA)

FIGURE 2-6 Hydrolysis of DF with warm water. SOURCE: Darryl Palmer, Office of the PMNSCM, "Hydrolysis of DF, Hydrolysis of QL," personal communication to George Parshall, December 24, 2001.

chemicals. If neutralization is chosen for use at PBNSF, a punch-drain-neutralize component would be built and would operate in the 2006-2007 period.

Technical Issues

As demonstrated by the campaigns at APG, the neutralization of DF and QL should be considered a mature technology, especially in light of the Army's experience in destroying large quantities of GB by a similar process. The neutralization of DF and QL can be carried out in ordinary commercial stirred-tank reactors, which are available in almost any size appropriate for the task.

The neutralents may require treatment before ultimate disposal because of residual toxicity and because they contain methylphosphonic acid (MPA) derivatives that are CWC Schedule 2 precursors. (DF and QL are specifically listed in the treaty as Schedule 1 precursors.)

Regulatory Approval and Permitting Issues

In general, neutralization seems to be regarded by regulators and the public as a relatively safe technology. Regulatory approval and permitting should be much easier than for technologies employing high temperatures or pressures. However, a large amount of secondary waste is produced by neutralization processes, which requires further treatment prior to ultimate disposal.

Public Concerns

The mild reaction conditions and minimal emissions from neutralization have generally led to its acceptance by the public as a demilitarization technology.

Open Burning/Open Detonation

OB/OD is a traditional process that has been used to destroy waste munitions, both conventional and chemical, for many years.

Description

Military explosive ordnance disposal personnel are taught to use OB/OD to dispose of individual chemical rounds by countercharging the round with 5 pounds of explosive to every pound of chemical fill (OD) or by remotely opening and burning the chemical fill (OB). Using OD, the nearly instantaneous heat and pressure of the detonation of the surrounding explosives is believed to destroy most of the chemical fill, rendering the munition harmless in an inexpensive and expeditious manner. OB can be used in a field process whereby a munition is positioned on a large burn pile and is opened explosively from a distance after remote initiation of the burn.

Status

OB/OD remains a useful disposal option. For example, a military team working with explosive ordnance devices may recommend using OD to dispose of chemical munitions that are believed to be in a dangerous condition, either because the fuze is armed and shock sensitive or because the munition has seriously deteriorated. Using OD, the munition need not be moved because it can be disposed of in place.

Technical Issues

OB/OD of non-stockpile munitions is simple, inexpensive, and expedient. No secondary waste streams requiring further treatment and disposal are produced unless soil and fragments surrounding the OB/OD site are removed. Using the OB/OD for disposal of chemical munitions is based, of course, on the assumption that the area can withstand a significant high-order detonation and that no personnel or property are located in the downwind hazard area.

At the same time, OB/OD has several disadvantages. The process is noisy, and the high ratio of explosives to agent required for OD and the large fire pit required for OB make OB/OD impractical for the treatment of very large munitions containing large quantities of agent. There is little information on how much of the chemical fill of a weapon remains after OB/OD disposal and how much of it is released into the atmosphere as vapors or particulates. In addition, the disposal of conventional high-explosive munitions by OB/OD has resulted in measurable environmental contamination from high-explosive residues. At some sites, such as the Massachusetts Military Reservation, long-term, heavy use of military high explosives has contaminated the groundwater.

Regulatory Approval and Permitting Issues

Today's environmental regulations make disposal of non-stockpile munitions by OB/OD difficult or impossible, except perhaps under extreme (e.g., warfare) conditions. For example, the Army initially proposed OD to treat the GB bomblets found at RMA, but this option was strongly opposed by state of Colorado regulators. The Army lacks critical information about environmental releases of agent and other hazardous materials during OB/OD, and any attempt to permit OB/OD for routine or emergency disposal operations would likely face strong regulator and public opposition.

Public Concerns

Public concerns about OB/OD include the noise from repeated detonations and uncertainties about the types and quantities of toxic materials released in the process. Strong public opposition to future use of OB/OD for treatment of CWM can be expected.

Integrated Ballistic Tent and Foam System

The Integrated Ballistic Tent and Foam System (hereinafter called the tent-and-foam system) allows partially contained blow-in-place disposal of chemical munitions (as large as 8-inch projectiles) that cannot be inserted into a mobile disposal system (e.g., the EDS) or transported to a disposal facility because they are badly deteriorated or are extremely susceptible to accidental detonation. The inability to safely move the chemical munition is expected to be rare but very problematic.

Description

The tent-and-foam system consists of an inner tent filled with blast suppressive foam, an outer ballistic tent, and an integrated air pollution control system. The tent-and-foam system is designed to contain the blast, fragments, and toxic emissions resulting from the detonation of a chemical munition using 5 pounds of donor explosive for every pound of chemical agent fill.

The inner primary tent, constructed of a ballistic material designed to contain blast overpressure and fragments from the detonation, is 6 feet on each side at the bottom, 2 feet on each side on the top, and 4 feet high. The primary tent is supported by flexible fiberglass rods and a fill sock that allows the interior of the tent to be filled with a combination of blast-suppressive foam and chemical decontaminating agent to contain the blast and neutralize the released chemical agent.

The outer secondary tent is made of Kevlar or another ballistic material and is supported by an air-rib system. It is placed over the primary tent to provide vapor containment of chemical agent and is connected with flexible stainless steel hoses to the air pollution control system. A slight negative pressure is maintained within the secondary tent prior to detonation of the chemical munition.

The air pollution control system has an acid gas wet scrubber and a charcoal filtration unit, transported on one trailer. The gases flow from the secondary tent through the acid gas scrubber and demister and then through the carbon high-efficiency gas adsorbers and a high-efficiency particulate adsorber. Although some gases are expected to escape the system, most of the toxic chemical agent fill of the munition will be either consumed during the detonation or captured in the air pollution control system.

Most of the solid waste, consisting of the fragmented remains of the munition, will remain within the footprint of the secondary tent, but some is expected to escape randomly.

It is expected that one munition can be disposed of in one day using the tent-and-foam system, which cannot be reused.

Status

The tent-and-foam system is in the final stages of development; however, it has not been tested with simulants or chemical agent. When the Army completes its testing, the system will likely replace OB/OD for destruction of NSCWM in nonwarfare conditions.

Technical Issues

While the committee anticipates that the tent-and-foam system may reduce the amount of toxic residue and confine the spread of contamination to a smaller footprint compared with OB/OD, test data are required to ensure that it reduces contamination to an acceptable level.

Regulatory Approval and Permitting

It is highly unlikely that regulators will permit the use of the tent-and-foam system until the amount of released agent can be determined. Testing to determine the amount of agent and other emissions released from the system during disposal of chemical munitions was scheduled to begin in

March 2002 using simulants in the Prototype Detonation Test and Destruction Facility in the Edgewood area of Aberdeen Proving Ground.

Public Concerns

Public concerns about the tent-and-foam system are likely to be similar to those about OD. However, the fact that the blast is partially contained and that the foam and the air pollution control system are expected to neutralize or capture much of the agent that survives the initial blast may make the tent-and-foam system a more attractive option.

Multiple-Round Containers

Several mobile treatment systems discussed earlier in this chapter can be dispatched to the site of an NSCWM discovery to perform on-site treatment of the item. An alternative to the use of mobile systems is to overpack the chemical waste to be treated in a multiple-round container (MRC) and transport it to an off-site facility for treatment or storage.

MRCs are a family of six overpack containers designed to allow the safe transport of NSCWM. The dimensions and capabilities of MRCs are shown in Table 2-5. MRCs are not designed or intended to contain the accidental detonation of a chemical munition. They should be considered only as overpack containers for containment of internal leaking and for protection of their contents during accidents or rough handling. Any chemical munitions to be transported in an MRC must first be determined to be safe, that is, explosion-proof, to transport by military ordnance experts.

Fragile items such as CAIS may also be transported. Individual CAIS vials are usually placed in cardboard mailing tubes and then packed in the MRC with vermiculite to act as a cushioning material. Complete PIGs can also be transported in the MRC 12x56, which was designed with this purpose in mind.

Neutralents from treatment of NSCWM are generally considered hazardous wastes. Their transport does not require MRCs; instead, they may be transported in DOT-approved containers according to the regulations controlling hazardous wastes.

Status

The MRCs were specifically designed, tested, and fielded for transport of CWM and exceed the United Nations' performance-oriented packaging requirements. They have been fielded and are currently in use, having been approved by the Department of Defense (DOD) and DOT for the storage and transportation of RCWM in the public domain.

Technical Issues

No outstanding technical issues remain to be resolved with respect to the use of MRCs.

Regulatory Approval and Permitting Issues

Transportation of NSCWM is heavily regulated (see discussion in Appendix G). State regulators have not identified any deficiencies in the design or use of MRCs.

In some cases, such as when NSCWM is discovered in an urban area where it is not feasible to bring in a mobile treatment system such as the EDS—e.g., the Spring Valley site in Washington, D.C. (U.S. Army, 2002a)—transportation is the only option. Even here, however, the Army faces great challenges in identifying a site willing to receive the NSCWM. PMNSCM has indicated that it has "worn out the welcome mat" at sites such as APG and PBA, which accepted such shipments in the past. As a result, PMNSCM no longer considers the transportation of NSCWM to be an option for routine use; it is considered as an option only in extreme or emergency circumstances.

TABLE 2-5 Multiple-Round Containers and Their Contents

MRC	Designed to Contain
MRC 7x27	Items up to 7 in. in diameter, 27 in. long, and weighing up to 100 lb
MRC 9x41	Items up to 9 in. in diameter, 41 in. long, and weighing up to 200 lb
MRC 12x56	Items up to 12 in. in diameter, 56 in. long, and weighing up to 200 lb
MRC 16.5x5.5	Items up to 16.5 in. in diameter, 5.5 in. long, and weighing up to 50 lb
MRC 21x79	Items up to 21 in. in diameter, 79 in. long, and weighing up to 1,000 lb
MRC 30x40	Items up to 30 in. in diameter, 40 in. long, and weighing up to 850 lb

SOURCE: Provided to the committee by Thomas Huff, Office of the PMNSCM, October 17, 2001.

Public Concerns

There tends to be strong opposition from local public groups to the transportation of hazardous materials such as NSCWM through their communities. There is also great reluctance on the part of public stakeholders at potential receiving sites to accept shipments of NSCWM (especially from out of state) lest these sites become a dumping ground for such materiel in the future.

Finding 2-13a. MRCs will be an extremely valuable asset if the Army pursues the transportation of small quantities of recovered CAIS to the RRS or commercial disposal facilities or the disposal of NSCWM at stockpile disposal facilities.

Finding 2-13b. In some cases, transportation of NSCWM from the site of its recovery to a secure location will be the most appropriate option.

Recommendation 2-13. The Army should work with state regulators at specific installations to allow transport of NSCWM from off-site locations for storage or treatment under extreme or emergency circumstances.

3

Application of the Non-Stockpile Treatment Systems to the NSCWM Inventory

INTRODUCTION

Chapter 1 described the kinds of challenges that the Army faces in disposing of NSCWM. Chapter 2 focused on defining the technologies and systems available to the Product Manager Non-Stockpile Chemical Materiel (PMNSCM) for destroying the non-stockpile inventory. In this chapter the committee matches the systems with the materiel or munitions that have been or may be recovered to determine if there are any gaps in the current program. Technologies or systems for treating secondary wastes (energetics, neutralents, rinsates, solids, etc.) are also matched. The committee then reviews the nature of the NSCWM treatment categories by comparing type, quantity, condition, and location with the available systems or technologies. Because this evaluation of the candidate systems and technologies relative to the Army's needs involves an exercise of judgment by the committee, few outside sources can be cited. Lastly, findings and recommendations are provided,[1] followed by a brief discussion of future program needs that the Army should address for the successful accomplishment of its mission.

COMPARISON OF CANDIDATE TECHNOLOGIES AND NEEDS

Table 3-1 compares the technologies and systems with the materiel and munitions that constitute the non-stockpile inventory. As shown in the table, no tools are available for very large bombs, some of which have recently been discovered outside the United States. Most of the remaining inventory items have several alternative means for disposal, and most of the available technologies have several applications.

Table 3-2 then looks at the available technologies for treating secondary wastes. There are multiple choices available for the treatment of agent, neutralent, and rinsate. Not all of these technologies are needed to provide adequate coverage.

The committee then matched non-stockpile CWM treatment requirements to the CWM treatment systems currently available and under development. In this analysis the following 10 operational CWM treatment categories were considered:

1. CAIS PIGs[2]
2. individual CAIS vials and bottles
3. small quantities of small munitions
4. chemical agent in bulk containers
5. binary chemical weapons materiel components
6. unstable explosive munitions that cannot be moved
7. secondary liquid waste streams
8. large quantities of NSCWM items currently in storage
9. large NSCWM items
10. large quantities of not-yet-recovered small munitions

These operational CWM treatment categories are discussed in the following sections, and the suitable treatment technologies for each category are identified. The commit-

---

[1] In some cases, the relevant findings and recommendations were stated in Chapter 2, and the reader is referred to the appropriate sections of that chapter.

[2] PIGs are metal canisters with packing material designed to protect CAIS sets during transport.

TABLE 3-1 Match of Primary Technologies and Systems to Items in the Non-Stockpile Inventory

Candidate Primary Technologies and Systems	CAIS Vials	CAIS in PIG	Binary Canisters	Fuzes and Bursters	Mortars	Projectiles	Small Bombs and Bomblets	Livens Projectiles	Traktor Rockets	Mines	Spray Tanks	Rocket Warheads	Ton Containers	Chemical Vials	Buckets and Drums	Bottles	500-1,000 lb Bombs
SCANS	T_4																
RRS	T	T															
Donovan blast chamber	P			P	P	P	P	P		P		P		P		P	
EDS	T		T	T	T	T	T	T	T_2	T	T_1	T	T_1	T		T	
Drill and drain and neutralization			T	T	T	T	T	T	T	T	T	T	T				
Commercial incineration	T		T														
Stockpile incineration	T		T	T	T_5	T_5	T_5	T_5	T_5	T_5	T_5						
Other Army facilities (including CTF and CAMDS)			T								T	T	T	T	T	T	
Tent and foam				T	T	T	T	T		T		T					
Plasma arc	T	T	T	T	T_3	T_3	T_3	T_3	T_3	T_3	T_3	T_3	T_3	T_3	T_3	T_3	
Bulk neutralization			T		T	T	T	T			T		T	T	T	T	
Drill-through valve																	P
Batch SCWO-EDS	Insufficiently mature to evaluate																
OB/OD	No obvious role, so not considered by the committee																

NOTE: Unless indicated by a P or a T, materiel and munitions are not found to fit candidate primary technologies and systems. Legend is as follows:
P, potential fit;
T, technical fit (see text for specific exceptions);
T_1, technical fit if agent is repackaged into smaller containers (e.g., DOT bottles);
T_2, may be a technical fit if the motor is removed;
T_3, technical fit if larger solids reduced in size;
T_4, still in design phase; and
T_5, technical fit if compatible with stockpile demilitarization machines or if demilitarization machines can be modified.
SOURCE: Compiled by the committee.

TABLE 3-2 Focus of Secondary Technologies

Candidate Secondary Technology	Energetics	Neutralents	Rinsates	Solids
Plasma arc		X	X	X
Commercial incineration		X		X
Chemical oxidation		X	X	
Stockpile incineration		X		X
Wet air oxidation		X	X	
Caustic hydrolysis	X	X		
POTW/FOTW			X	
Peroxide/TiO$_2$/UV			X	
Explosive containment chambers	X			
Metal parts furnaces				X

SOURCE: Compiled by the committee.

tee also notes where there are no existing technologies for an identified category.

CAIS PIGs

Finds of complete CAIS PIGs have added a technical requirement for opening the PIG container to access the individual CAIS vials and identify their content prior to treatment. This additional requirement prevents using the Explosive Destruction System (EDS) or Single CAIS Accessing and Neutralization System (SCANS).

The RRS is an existing system that was designed specifically for handling and treating complete CAIS PIGs. Either the RRS can be brought to the location of the CAIS PIG or a PIG can be shipped to the location of the RRS in the multiple-round containers (MRCs) developed for this purpose. In both cases, the PIGs will be opened, the CAIS vials neutralized in the RRS, and the resulting neutralent treated at a commercial TSDF, as planned by the Army. Neutralent could also be treated in stockpile facilities if regulator and public approval can be obtained. CAIS vials containing industrial chemicals are separated from those containing agent and sent to a TSDF for disposal.

As with small-quantity CAIS finds, MRCs could conceivably be used to ship CAIS PIGs to a stockpile incineration facility or one of the planned non-stockpile treatment facilities; however, the receiving state must be willing to accept the materiel to be received and treated (see Appendix G for a discussion of transportation issues).

Because there is a wide range of suitable treatment options for this NSCWM category, the committee believes that PMNSCM has adequate options to handle and treat complete CAIS PIGs.

Individual CAIS Vials and Bottles

This category is defined as a find of CAIS vials or bottles not containerized in a PIG. Loose CAIS vials or bottles have been found at military training areas (e.g., Fort Ord, California, and Fort Meade, Maryland) during remediation projects for other contaminants, and additional finds are expected in the future. These vials and bottles are suspected to be discarded CAIS that remained after training exercises were completed.

There are three existing systems capable of adequately handling these finds. The first is neutralization in an RRS, followed by disposal of the resulting neutralent and solid waste. A second option is the explosive opening of loose CAIS items in an EDS, followed by neutralization of the products in the EDS and disposal of the neutralent and solid waste.[3] The third option is transportation of the loose CAIS items in an MRC to the location of a stockpile or non-stockpile facility, mobile system, or commercial TSDF for disposal.

Two developmental systems could also be used. First, the SCANS system for on-site disposal of individual CAIS vials or bottles is under development specifically for this category. The benefit of SCANS is that it is designed to be an efficient treatment system that can be used if an RRS or EDS is not available for relocation to the site of the CAIS find and if transportation in an MRC is not allowed by the receiving state. SCANS may be a worthwhile alternative disposal option that can save the expense and time of transporting an RRS or EDS to the site. Second, if the development and permitting of the Donovan blast chamber (DBC) are completed, this system could be used for the disposal of loose CAIS, with the DBC transported to the CAIS or the CAIS

---

[3]Some components of CAIS sets, such as lewisite, charcoal, and neat chloroform, have never been introduced into the EDS, and it is possible that treatment procedures might have to be modified to accommodate these materials. Testing would need to verify that treatment goals can be met. However, the committee does not anticipate that any serious difficulties would be encountered in treating CAIS vials and bottles in the EDS.

transported to the DBC in an MRC, although the latter might not be cost-effective.

The committee believes that PMNSCM has adequate treatment options for this category.

Small Quantities of Small Munitions

For the purposes of this analysis, small munitions are defined as those that can be efficiently processed in the EDS-1 or EDS-2 (the latter will be capable of destroying chemical munitions as large as World War I-type 8-inch or a modern 155-mm projectile, including the necessary shaped charges). For this analysis, a small quantity is defined as the amount that can be treated efficiently in a single EDS.

The quantity is specified as "small" because with only six additional EDSs planned for production, it may not be appropriate to tie up the existing EDS capability (throughput is currently 2 days per CW munition) for an extended time at only one or two sites with large quantities of CW munitions, to the detriment of other sites with munitions awaiting disposal. However, the EDS has been demonstrated to be an appropriate method for disposing of small quantities of small CW munitions: 10 GB-filled bomblets were treated at the former Rocky Mountain Arsenal.

Also, MRCs can be used to transport small quantities of small munitions to the current location of an EDS, instead of moving the EDS to every small find. It might also be possible to transport such finds in MRCs to one of the planned non-stockpile disposal facilities if regulatory approvals for transportation and treatment can be obtained.

The DBC is another disposal system suitable for small quantities of small munitions. It is currently being tested and evaluated for disposing of CWM in Belgium. The DBC offers an alternative treatment option for neat chemical agents that avoids the addition of chemical reagents and the generation of associated liquid waste streams. However, increased regulatory approval difficulties are anticipated for the DBC, because its agent destruction efficiency is only 99 percent[4] and there is potential for worker exposure. Thus, the EDS is currently the more desirable disposal system in this category. Nevertheless, the DBC might be operable in a manner that protects workers and the community, particularly if it is a component of a multistage treatment system. For example, if the residuals from several days of operation of the DBC (which are expected to contain about 1 percent chemical agent) are subject to further treatment (e.g., neutralization or incineration), the destruction and removal efficiency of the entire treatment system may exceed 99.9999 percent. Thus, the DBC could be used if these issues, including gas/vapor containment and contaminated particulates issues, can be resolved and regulatory approvals can be obtained.

The committee believes that PMNSCM has an adequately wide range of disposal options, existing or planned, that are suitable for this disposal category.

Chemical Agent in Bulk Containers

Because all known non-stockpile ton containers are located at or adjacent to stockpile sites (U.S. Army, 1996), the committee recommends that the stockpile facilities should be used to process them to the extent possible (Chapter 2). Use of the stockpile facilities would destroy the neat agent directly, avoiding the addition of chemical reagents and the generation of associated liquid waste streams. Two sites will require special arrangements (Appendix C). At the Bluegrass Army Depot, regardless of the technology chosen, there is a GB-filled ton container that will have to be drained into smaller containers, because the Bluegrass Chemical Disposal Facility (BGCDF) will not have facilities for handling ton containers, although it will be able to destroy GB. At Pine Bluff Arsenal (PBA), two GB-filled ton containers can be handled in the Pine Bluff Chemical Disposal Facility (PBCDF) without equipment modification. However, some of the more than 4,000 empty ton containers at Pine Bluff may have contained lewisite, and these must be decontaminated in a special non-stockpile facility because the PBCDF incinerator will not be equipped to process arsenic-containing materiel.

Smaller bulk containers (vials, bottles, buckets, and drums) are also located primarily at stockpile sites and generally appear suitable for disposal in stockpile incinerators if these facilities can be permitted to destroy the agents in the smaller bulk containers and if the agents can be accessed by the demilitarization machines at the stockpile facilities. In some instances, modest equipment changes will be required as well as permit modifications (Appendix D). Chemical samples at PBA and the Aberdeen Proving Ground (APG) cannot be disposed of in those CDFs, but suitable options are available in the planned facilities (PBNSF and MAPS) that include drill-and-drain and bulk neutralization capabilities. The Chemical Agent Munitions Disposal System (CAMDS) facility at Tooele also has facilities for disposal of nonstandard sample containers stored at Deseret Chemical Depot (DCD) and Dugway Proving Ground (DPG).

The committee believes that the disposal of bulk containers of chemical agent can be addressed through the use of existing or proposed facilities (i.e., the stockpile facilities, CAMDS, PBNSF, MAPS), providing that the necessary equipment and permit modifications are made, and assuming all transportation issues can be addressed (see Appendix G).

Binary Chemical Warfare Materiel Components

The only known inventories of the phosphorus-containing binary CWM components DF and QL are stored at PBA in canisters and drums. Either a plasma arc system or the

---
[4]Herbert C. De Bisschop, personal communication to G.W. Parshall in an interview at the Belgian Royal Military Academy on July 25, 2001.

PBCDF incinerator would destroy the neat binary precursors directly, thus avoiding the addition of chemical reagents and the generation of associated liquid waste streams. Several options for direct destruction of these compounds appear attractive but may not be feasible. Burning them at a commercial incinerator is unlikely because the TSDFs capable of handling the fluoride content of DF may not wish to accept them due to concerns about security, toxicity, and handling hazards.[5] Currently, the Army plans to destroy the binary CWM components using a plasma arc system that is being proposed as part of the Pine Bluff Non-Stockpile Facility (PBNSF). Disposal in the PBCDF incinerator would be very attractive if the permit could be modified to include binary CWM destruction. The schedule for operation of the incinerator may not, however, allow its use for non-stockpile purposes before the 2007 CWC schedule date (Appendix C).

Neutralization on-site is an attractive alternative for destruction of this materiel if direct destruction by a high-temperature process is not feasible. As described in Chapter 2, neutralization of DF and QL by water hydrolysis was demonstrated in a campaign at APG in 1997. However, this procedure produces neutralents that contain large amounts of CWC Schedule 2 precursor compounds. Therefore, it may be necessary to incinerate the neutralents or to develop an oxidative posttreatment system to destroy these compounds prior to final disposal. The DF neutralent will also contain fluoride, which may limit the number of commercial TSDFs capable of destroying it. However, one TSDF operator has indicated that his facility frequently burns high-fluoride solutions.[6] Alternatively, the fluoride content of oxidized neutralent could be immobilized by treatment with calcium hydroxide (slaked lime) in preparation for final disposal.

All in all, the committee believes that either the PBCDF incinerator or the proposed PBNSF plasma arc system could be used to efficiently dispose of the binary CWM components, providing the necessary equipment and permit modifications can be made. Destruction of the binary CWM components by the CWC deadline of April 2007 will be a challenge, however; the Army needs to incorporate realistic milestones for destruction of the binary CWM components into the operating schedule for the selected system.

Unstable Explosive Munitions That Cannot Be Moved

This category of CWM includes all munitions that are not suitable for movement into one of the existing or planned disposal systems because their fuzing has been determined to be extremely shock sensitive or because they are in such a severely deteriorated condition that movement could cause a leak.[7] A recovered CWM that is so unstable that it cannot be safely moved into a mobile chamber, such as the EDS, would be classified as an emergency.

Historically, open detonation has been used to destroy this type of recovered munition. More recently, however, the public and regulators generally have begun to consider open detonation of CWM unacceptable. One way to deal with such situations is to detonate the object in an enclosure such as the tent-and-foam system described in Chapter 2. The NSCMP has already performed some limited tests on this system to mitigate the environmental impact of open detonation.

Use of the tent-and-foam system will require locating the CW munition in an area that can withstand a high-order detonation or constructing engineering controls to mitigate the potential damage of a high-order detonation. However, this disposal scenario is expected to arise relatively infrequently and most likely would occur on a former CWM testing range, where a high-order detonation is more feasible.

The committee believes that development and testing of the tent-and-foam system should be completed to fulfill the needs of PMNSCM for this disposal category. In light of the expectation that the tent-and-foam system will reduce the amount of agent contamination released from the in-place disposal of a CW munition by detonation, there is no disposal scenario foreseen that would require conventional open detonation of CW munitions, except for expedient CWM disposal in wartime under battlefield conditions.

Secondary Liquid Waste Streams

Treatment systems such as the RRS and EDS that rely on chemical neutralization of agents produce secondary liquid waste streams of two types:

- neutralized agent (neutralent) waste streams consisting largely of organic solvents and agent neutralization by-products
- aqueous waste streams, including rinsates, washes, and brine solutions

The Army's plan for destroying these wastes involves collection of neutralent, rinsates, and washes, followed by treat-

---

[5]Christopher Ross, PMNSCM, briefing to the committee on November 9, 2001.

[6]Don Matter, manager of Safety-Kleen facility in Deer Park, Texas, personal communication to Joan Berkowitz, committee member, on January 30, 2002.

[7]Because of the possibility of finding CWM in a deteriorated condition, it is assumed by the committee that the recovery of CWM will be done under controlled conditions (e.g., working within protective negative-pressure shelters), by workers outfitted in the appropriate level of personal protective clothing. It is further assumed that the project work plan will provide guidelines to the specially trained workers determining whether or not a munition can be moved and how to handle and process the recovered munitions.

ment on-site or shipment to a public or federal treatment, storage, and disposal facility (TSDF) for final disposal. Disposal of these neutralents, rinsates, and washes would typically be by incineration. Off-site shipment could be an expedient option, provided that the Army is able to resolve a number of technical, regulatory, and public perception issues satisfactorily. Input from the interested public on the acceptability of the option must be obtained through open and objective discussion prior to implementation. TSDFs with suitable technologies for treating the neutralents may have to obtain permit modifications before they can accept the materials. Moreover, the Army will need to obtain long-term commitments from suitable TSDFs to accept the wastes.

If off-site shipment does not prove to be feasible for any reason, a development program will be required to demonstrate the viability of potential on-site treatment technologies. Plasma arc systems might be adaptable to destruction of secondary wastes from EDS and RRS neutralization. However, use of such high-temperature processes to destroy highly aqueous secondary wastes would be inefficient, although it may be expedient in some cases. If the aqueous waste liquids cannot be disposed of in publicly or federally owned treatment works (POTWs or FOTWs), further development work will be required to demonstrate alternatives. A previous report of this committee (NRC, 2001a) identified chemical oxidation or wet air oxidation as potentially promising, but as of November 2001 neither had been demonstrated for neutralents. A development program would need to address issues such as the following:

- materials of construction and corrosion
- DREs of specific by-products
- salts/solids management
- safety issues
- VOC and emission control

PMNSCM has undertaken a technology test program to test a large number of alternative technologies for destruction of these secondary waste streams. Some of the alternative technologies appear to be only marginally appropriate for the needs of the program. Previous reports of this committee (NRC, 2001a, 2001b) pointed out the limitations of biological treatments, electrochemical oxidation, gas-phase chemical reduction, solvated electron technology, and continuous SCWO technologies. The committee has recommended no further development of these technologies for treatment of neutralents and rinsates. In addition, the current analysis points out the need for more information on the scope of applicability of UV-catalyzed oxidation and calls into question the need to continue the development of batch SCWO processing. The potential applicability of UV oxidation is not well understood, and batch SCWO, when combined with explosive accessing in an EDS, is technologically immature. The committee believes that the Army's plan for destruction of secondary NSCWM liquid waste streams (discussed above) is sound, but that further development of plasma arc, chemical oxidation, and wet air oxidation should be conducted to handle cases where incineration of the wastes at a TSDF or disposal at a POTW are inappropriate or unacceptable to the local public (see earlier findings and recommendations 2-12a and 2-12b).

Large Quantities of NSCWM Items Currently in Storage

Pine Bluff Arsenal has the largest known non-stockpile inventory. It contains almost 70,000 items, including explosive and nonexplosive munitions with diverse chemical fills, binary agent precursors, CAIS, and bulk containers of chemical agent. If no extension of the CWC treaty deadline is sought, these items must be destroyed by April 29, 2007.

PMNSCM has plans for construction of a facility (PBNSF) designed to dispose of these items. It may also be used to dispose of other finds of CAIS or CW munitions if transportation and public acceptance issues can be resolved. As discussed previously, if transportation to PBNSF is not allowed, other disposal options, including the deployment of a mobile system with higher throughput than the EDS or the building of more facilities at the sites where NSCWM is discovered, may need to be implemented.

While the design for PBNSF that the Army has shared with the committee appears to be technically suitable for the treatment of NSCWM at Pine Bluff (see Appendix C), the task of destroying this quantity of NSCWM by 2007 is daunting given that PBNSF is not expected to be operational until 2006. As far as the committee can ascertain, the Army has not developed a realistic timetable for destruction of this NSCWM that is consistent with current treaty deadlines. The committee is concerned that without clear planning and extraordinary efforts, the treaty deadlines will almost certainly not be met.

Large NSCWM Items

There is currently no fully developed treatment system capable of handling munitions larger than those that can be processed in the EDS-2 and possibly the DBC. This limits the on-site disposal of CW munitions to 155-mm projectiles or 8-inch World War I chemical projectiles, because the amount of explosives contained in larger munitions exceeds the capability of the EDS and the amount of donor explosives required for destruction in the DBC exceeds its capabilities. Examples of large items include 500- and 1,000-lb chemical bombs.

Based on the latest assessment of buried non-stockpile munitions (U.S. Army, 1996), it is likely that this disposal scenario will exist in the United States in the future. Also, large chemical bombs (M-79 1,000-lb bombs and M-78

500-lb bombs) are known to exist in Panama (OPCW, 2001). PMNSCM currently does not have a fully developed and field-ready system capable of processing these large chemical bombs. This represents a potentially important gap in Army treatment capabilities.

One promising approach is the British-developed drill-through valve (DTV) system (U.S. Army, 2000a). The DTV is designed for large munitions with liquid (nongelled) agent fills (e.g., hydrogen cyanide, CG, or CK). The DTV uses a drill-seal-and-drain system mounted on a saddle that is attached to the munition by epoxy. After drilling, the access hole is used for removal of the chemical fill and the subsequent injection and removal of decontamination liquid to decontaminate the bomb interior. The DTV may prove difficult to use when agent is present in a thickened form, but perhaps it could be modified for this purpose. If a sealed access hole can be obtained, appropriate draining, pumping, and decontaminating procedures might be developed for these agents. However, complete containment and destruction of agent are difficult to achieve by this method, and significant development is necessary before the system could be permitted in the United States.

If the necessary permits can be obtained to test the DTV system for the disposal of large CW bombs already discovered in Panama, such testing could provide valuable data on its performance that would facilitate permitting in the United States.

Large Quantities of Not-Yet-Recovered Small Munitions

A large quantity of small munitions is defined as an amount in excess of the quantity that can be efficiently treated in an EDS. Although the EDS will be technically capable of handling a large number of small CW munitions, its relatively low throughput rate means that a disposal system with a higher throughput may be desirable. Examples of sites where large quantities of not-yet-recovered small munitions are known to be buried include Dugway Proving Ground, Utah; Rocky Mountain Arsenal, Colorado; and Redstone Arsenal, Alabama (U.S. Army, 1996).

Treatment options for large quantities of small munitions, in addition to a prolonged campaign in an EDS, are packing the munitions in MRCs and shipping them, perhaps across state lines, to a facility; dedicating multiple EDS units to the task; using a transportable system with a higher throughput; or constructing a dedicated treatment facility. Shipment of large quantities of CW munitions to a facility is not likely to be approved by the receiving state, making this option unlikely. Current Army planning limits EDS procurement to seven units by FY 2007. Unless enough units are procured to allow the flexibility of dispatching several to a given site while keeping on hand adequate units for emergency use or achieving disposal schedules at existing sites, this option will not be possible.

Remaining options for this category are construction of a facility or development of another transportable CWM treatment system with a higher throughput than the EDS. If the relatively high cost and large footprint of a facility are judged to be inappropriate for the site, a high-throughput transportable system may be attractive. Assuming that technical and regulatory barriers with the DBC, as discussed previously, can be overcome, its incorporation into the available CW munition treatment systems for large quantities of small munitions would be a valuable addition to the variety of treatment systems currently available to PMNSCM. With this option, secondary waste would require further treatment to destroy residual agent.

However, as stated previously, the DBC is not presently in the PMNSCM arsenal of disposal tools, and its evaluation thus far has been underwritten by the U.S. Army Corps of Engineers (USACE). Throughout this study, it became apparent that there was a division of responsibility within the Army for accomplishing its remediation mission. The NSCMP is a development program under PMCD and, until recently, within the acquisition organization of the Army. Beyond the responsibilities of PMNSCM exists the responsibilities of the USACE. USACE is responsible for remediation of formerly used defense sites (U.S. Army, 2002a), including both conventional and chemical ordnance. In that role, the USACE is apparently investigating technologies and methodologies for non-stockpile munitions treatment. With the exception of the DBC, the committee restricted itself to evaluating PMNSCM technologies and systems. The committee strongly believes, however, that an integrated approach to the problem of chemical weapons remediation would serve the Army well.

NSCWM TREATMENT CATEGORIES FOR WHICH AVAILABLE OR IN-PIPELINE TOOLS ARE ADEQUATE

There is at least a sufficient range—and often a wide range—of disposal or destruction options available or under development for the following seven categories of non-stockpile materiel:

CAIS PIGs

The RRS is a system that was designed and developed by PMNSCM specifically to handle and treat complete CAIS PIGs and large numbers of loose CAIS vials and bottles. The committee finds it an expensive but adequate treatment system for CAIS PIGs and large numbers of CAIS vials (see RRS discussion in Chapter 2).

Individual CAIS Vials and Bottles

PMNSCM is developing the Single CAIS Accessing and Neutralization System (SCANS) to treat individual CAIS

Small Quantities of Small Munitions

PMNSCM has developed the transportable Explosive Destruction System (EDS)[8] as the workhorse system for destruction of both explosively and nonexplosively configured munitions in the field. The EDS-1 prototype was recently deployed to Rocky Mountain Arsenal, where it successfully destroyed 10 sarin bomblets. Improved versions of the EDS-1, as well as a larger EDS-2, are currently in development. Once these developments are completed, this category appears to be well covered. The EDS appears to be sufficiently flexible that it might also be used in other NSCWM treatment categories (see EDS discussion in Chapter 2 and Recommendation 2-5).

Chemical Agents in Bulk Containers

The non-stockpile inventory includes numerous containers of chemical agents of various types and sizes that have accumulated over the years. In general, these are stored at stockpile sites. There are many treatment options available for these bulk containers; the most obvious is to use the stockpile chemical disposal facilities (CDFs), although modifications may be required and permit modifications may be difficult to obtain.

In addition to the stockpile facilities, two experimental facilities have long been used to destroy a variety of chemical agents by chemical neutralization: the Chemical Transfer Facility (CTF) at Aberdeen Proving Ground, Maryland, and the Chemical Agent and Munitions Destruction System (CAMDS) near Dugway Proving Ground, Utah. Although these are R&D facilities and therefore should not be used on a routine basis to destroy NSCWM, they might be considered as an option to destroy limited numbers of non-stockpile items that contain unusual chemical fills or that have a configuration that cannot be handled by other systems.

Further treatment options for non-stockpile bulk chemicals include direct destruction in a plasma arc system (see below) or even treatment in the EDS. With all of these options available, this category is well covered.

Finding 3-1a. The stockpile chemical disposal facilities (CDFs) are capable of disposing of some of the non-stockpile inventory, although some modification in munition and container accessing equipment, agent monitoring, and pollution abatement equipment may be required.

Finding 3-1b. Bulk chemicals stored in ton containers and other sample containers are located at stockpile sites. They can be destroyed in the stockpile facilities, non-stockpile facilities at APG and PBA, or the CAMDS facility at Tooele, Utah.

Recommendation 3-1. While recognizing that there are significant regulatory and public acceptability issues to resolve, the committee recommends that non-stockpile chemical materiel in bulk containers located at stockpile sites and suitable for destruction in chemical stockpile disposal facilities be destroyed in those facilities.

Binary Chemical Warfare Materiel Components

The entire non-stockpile inventory of binary CWM components is stored in canisters and drums at Pine Bluff Arsenal, a stockpile site. Options for treatment include destruction in the Pine Bluff Chemical Disposal Facility, direct destruction in a plasma arc system (see Finding and Recommendation 2-10), or chemical neutralization followed by oxidative posttreatment of the neutralents. The high concentration of fluorine in the binary CWM component DF raises concerns about corrosion in some treatment systems.

Finding 3-2. Neutralization of the binary precursors DF and QL is feasible but generates substantial quantities of liquid wastes that contain CWC Schedule 2 precursors subject to oversight and inspection by the Organization for the Prohibition of Chemical Weapons. Posttreatment to destroy these secondary wastes will be required.

Recommendation 3-2. Ideally, the binary precursors methylphosphonic difluoride (DF) and ethyl-2-diisopropylaminoethyl methylphosphonite (QL) stored at Pine Bluff Arsenal should be destroyed directly, either by burning in the Pine Bluff Chemical Destruction Facility incinerator or by plasma treatment. If these facilities cannot handle the fluorine-rich DF destruction products, the committee recommends that on-site neutralization followed by oxidative posttreatment of the neutralents be developed. The easiest posttreatment may be shipment to a commercial incinerator capable of dealing with high levels of fluorine.

Unstable Explosive Munitions That Cannot Be Moved

Open burning/open detonation (OB/OD) was the traditional method for disposing of unstable munitions, including chemical munitions, but it is no longer considered acceptable by regulators except under emergency circumstances. PMNSCM has been exploring an alternative to OB/OD

---

[8]The EDS was originally developed to destroy non-stockpile items that were deemed to be too unstable for transport or long-term storage; however, it can also be used to treat limited numbers of stable chemical munitions, with or without explosive components.

called the tent-and-foam system, which provides for partially contained detonation of unstable munitions.

Finding 3-3. Unstable munitions discovered in environmentally sensitive or populated areas present a challenge because current technologies such as open burning/open detonation (OB/OD) are inappropriate.

Recommendation 3-3. The Army should complete the development and testing of the tent-and-foam system for controlling on-site detonation of unstable munitions.

Secondary Liquid Waste Streams

There appear to be a number of viable options for treatment of secondary liquid waste streams from systems such as the RRS, SCANS, and EDS, although further development work will be required (see the discussion of plasma arc systems, chemical oxidation, and wet air oxidation of neutralents and rinsates in Chapter 2 and Finding and Recommendation 2-11).

NSCWM TREATMENT CATEGORIES FOR WHICH SIGNIFICANT ADDITIONAL INVESTMENT AND PLANNING ARE NEEDED

For the following three categories, the committee judges the treatment options that are available or in the pipeline to be insufficient to permit the non-stockpile program to meet its goals. Additional investment or planning efforts are needed.

Large Quantities of NSCWM Items Currently in Storage

Some 85 percent of all recovered NSCWM in the United States is stored at Pine Bluff Arsenal. The Army has designed the Pine Bluff Non-Stockpile Facility (PBNSF) to destroy the almost 70,000 items stored there, but the facility is not expected to be operational until 2006. As far as the committee can ascertain, the Army has not developed a realistic timetable for destruction of this quantity of NSCWM that is consistent with current treaty deadlines. The committee is concerned that without clear planning and extraordinary efforts, the 2007 treaty deadline will almost certainly not be met (see PBNSF discussion and Finding and Recommendation 2-2 in Chapter 2).

Large NSCWM Items

Disposal of chemical projectiles larger than 155-mm and large (500- or 1,000-lb) bombs presents a special challenge for the non-stockpile program. Although such munitions are rarely recovered in the United States, they have been recovered as a result of U.S. activities in at least one foreign country, and it seems likely they will be found on U.S. soil in the future.

Finding 3-4. Large munitions, such as some chemical bombs and some chemical projectiles, cannot be treated in any of the planned or existing non-stockpile treatment systems because either the size of the munition or the amount of explosive exceeds the capacity of the treatment system. PMNSCM is investigating the suitability of the British drill-through valve (DTV) system, or some variant of that system, for use in accessing the chemical fill of large munitions.

Recommendation 3-4. PMNSCM should develop a strategy for treating chemical bombs and projectiles that are too large for treatment in the EDS, in the DBC (if successfully demonstrated), or in planned facilities. One option is to test the British drill-through valve (DTV) system, modify it if necessary, and prepare it for use on existing large NSCWM items and other such items that may be found in the future.

Large Quantities of Not-Yet-Recovered Small Munitions

Sites at which thousands of NSCWM items are believed to be buried present a special challenge to the non-stockpile program. Examples of such sites include Deseret Chemical Depot, Utah; Rocky Mountain Arsenal, Colorado; and Redstone Arsenal, Alabama. Use of one or even a few EDS units would be inefficient given their relatively low throughput capacity (currently one munition every 2 days). At present, the Army's only option for cleaning up such a site would be the construction of a facility, such as MAPS or PBNSF. However, such facilities are expensive and have a large environmental footprint. A transportable treatment system with a high throughput would be highly desirable to treat this category of NSCWM (see discussion of the Donovan blast chamber and Finding and Recommendation 2-9 in Chapter 2).[9]

DEVELOPING NEW SYSTEMS FOR NEW FINDS

There are large numbers of NSCWM items that are presently buried and that are likely to require removal and treat-

---

[9]While the committee believes that future non-stockpile facilities at large-quantity sites might utilize multiple mobile units such as the EDS and DBC operated in parallel (e.g., Finding and Recommendation 2-5), this concept is not yet included in facilities such as MAPS and PBNSF, which at this writing were under construction and in final design, respectively.

ment in the future (U.S. Army, 1996). The post-2007 period is approaching rapidly, and the transition from the destruction of stored NSCWM to sites that contain buried materiel must be addressed. This section discusses the types of criteria and guidelines the Army might consider in selecting treatment systems and technologies for not-yet-recovered CWM. The committee notes that it does not purport to recommend recovery and treatment of NSCWM items in all cases. In some cases, it may be preferable to leave the munitions buried and to institute institutional and other controls to protect human health and the environment. This section merely discusses treatment options should the decision be made to excavate and treat buried munitions.

Since buried CWM is known to exist at many sites (both current and former military facilities), the transition should include an assessment to ensure that these sites have treatment facilities that are adequate to treat the type and volume of buried CWM. Although the pressure of meeting the treaty deadline does not exist for buried NSCWM items, the Army still needs to set a reasonable schedule for the eventual destruction of this buried materiel. It is likely that the removal of buried CWM from the ground prior to destruction will pose the greatest risk, so the Army must have in place sufficient measures to ensure that human health is protected during removal operations.

The specific configuration of treatment systems at the existing non-stockpile mobile and planned facilities was based on a set of internal criteria and guidelines set forth by PMNSCM, but the criteria and guidelines were not provided to the public before decisions were made. As a result, some state regulators and some members of the public have indicated that the criteria for selecting technologies and treatment systems have been neither apparent nor documented. Although the committee believes that, overall, the treatment systems selected were reasonable and scientifically supportable, the administrative process might be improved if a general treatment system guidance document were developed and used in selecting the appropriate technology for a given site.

The choice of specific technologies or systems for a particular site depends on a range of factors, including the number of items, type of materiel, need for expeditious destruction, and proximity of existing stockpile or commercial treatment systems. No one set of treatment and disposal components can apply to all locations. At a new location, the Army must evaluate the ability of various alternative components to address the unique array of materiel at that particular location.

Finding 3-5. The Army has focused primarily on recovered chemical warfare materiel and its responsibilities for destroying this materiel by 2007 in accordance with the CWC. Relatively little emphasis has been placed on sites where significant quantities of NSCWM remain buried.

Recommendation 3-5. The Army should address the post-2007 period to ensure the smooth transition from destruction of stored NSCWM to that of buried NSCWM. Care should be taken to ensure the adequacy of treatment facilities for the type and volume of buried NSWM and that measures are in place for the protection of human health during removal operations.

4

Regulatory Approval and Permitting Issues

Before treatment technologies can be deployed to destroy non-stockpile chemical warfare materiel (NSCWM), regulators and the public must be satisfied that these operations will be carried out within the regulatory and legal framework established for protection of human health and the environment. The regulatory approval and permitting (RAP) process determines the choice of treatment technologies and the requirements they must meet and provides opportunities for public involvement in the decision-making process. The Non-Stockpile Chemical Materiel Product's (NSCMP's) RAP experience to date demonstrates that difficulties in resolving RAP issues can result in extensive delays and substantially increased costs for NSCWM treatment. An efficient and noncontentious RAP process can result in the timely destruction of NSCWM, in accordance with the Chemical Weapons Convention (CWC) schedule.

This chapter focuses on RAP for waste management, primarily that required under the Resource Conservation and Recovery Act (RCRA) but also, to a lesser extent, that conducted under the authority of the Comprehensive Environmental Response, Compensation, and Liability Act (CERCLA). Many of the principles discussed herein are also applicable to RAP under other environmental laws, including the Clean Air Act, the Safe Drinking Water Act, the Clean Water Act, and state-specific environmental legislation.

This chapter begins with a review of the Army's prior RAP experience for its various treatment systems and deployment actions. The committee then examines key issues that must be resolved to facilitate the RAP process in the future. Finally, the committee offers findings and recommendations to resolve these issues. Appendix F provides background on key regulatory provisions and a review of various RAP mechanisms that may be used for the destruction of NSCWM in accordance with regulatory requirements.

THE ARMY'S RAP EXPERIENCE

Prior experience can be summarized by examining permitting for various deployments of non-stockpile treatment technologies, including the MMD-1 (munitions management device) system, the RRS (rapid response system), the Spring Valley (Washington, D.C.) action, and the deployment of the explosive destruction system (EDS) to Rocky Mountain Arsenal (RMA) to destroy GB bomblets (NRC, 2001b). RAP efforts for the munitions assessment and processing system (MAPS) and Pine Bluff Non-Stockpile Facility (PBNSF) facilities are also reviewed.

Munitions Management Device

The Munitions Management Device was a trailer-mounted system developed by PMNSCM for the non-explosive destruction of NSCWM by drill-and-drain plus neutralization technology. The MMD-1 (a neutralization technology) was permitted using a RCRA Research, Development, and Demonstration (RD&D) permit at Dugway Proving Ground, Utah. It took more than 5 years to issue the RD&D permit, and even then, the initial permit was limited to the treatment of phosgene. Treatment of agent in the MMD-1 was still under negotiation when the Army cancelled the MMD program in March 2001. The Utah Department of Solid and Hazardous Waste (DSHW) and the Army were both frustrated by the experience. The DSHW experienced many problems in evaluating permit application documentation and negotiating permit requirements with the Army. The Army experienced numerous and changing demands for data and information from the state. The process may have been complicated by the MMD-1 design, which evolved considerably over time and which raised issues different from those raised by the stockpile incinerator facility at the neighboring Deseret Chemical Depot (DCD). These problems contributed to the Army's decision to cancel the MMD program. The EDS device was initially tested at Porton Down, in the United Kingdom, and the PMNSCM plans to initially test the EDS-2 device there as well.[1]

Rapid Response System

The RRS device was operated at DCD, Utah, under a standard RCRA permit. Approval of the RRS permit took more than 3 years, with the RRS mission at DCD having since been completed successfully. As had happened with the MMD-1, the DSHW and the Army were equally frustrated over the RRS permitting process, although the RCRA permit for it was easier to obtain than that for the MMD-1. While there are probably several reasons for this difference, the main ones would be the greater complexity of the MMD-1 system and the larger quantity of neat chemical agent to be handled.

Spring Valley, Washington, D.C.

In 1993, World War I era mustard and other munitions were discovered in the Washington, D.C., neighborhood of Spring Valley (U.S. Army, 2002b). The munitions were buried at the former American University Experiment Station, a World War I era military site used for the development of chemical warfare items. As a formerly used defense site (FUDS), Spring Valley is not part of a facility or installation. There was therefore no overarching environmental regulatory structure under RCRA or CERCLA at the time of the FUDS discovery. The action was processed using CERCLA removal authority in consultation with the Army and regulators from the Environmental Protection Agency (EPA) Region III and the District of Columbia. The response was initiated in 1994 and is ongoing.[2] The RAP at Spring Valley did not involve approval or deployment of treatment technologies. Most of the recovered items were transported to Pine Bluff Arsenal for storage and eventual treatment.

Rocky Mountain Arsenal, Colorado

At RMA, Colorado, the EDS was used to destroy 10 GB (sarin) bomblets. The first stage of the EDS operation was carried out through a state equivalent of EPA's imminent and substantial endangerment order (RCRA §7003), issued by the state of Colorado. RMA is listed on the CERCLA National Priorities List and has an ongoing remedial program, approved through a CERCLA record of decision. The Army used the existing structure under the RMA's CERCLA remedial program to destroy the sarin bomblets using the EDS; it met the requirements of the state's RCRA order as a CERCLA emergency removal action.[3] Doing so enabled the Army to avoid the long RAP delays experienced in Utah with RCRA permitting of the RRS and MMD-1. The Army hopes that the CERCLA removal approach can be a model for the RAP process at future sites to which the EDS might be deployed.[4] However, the CERCLA emergency removal action RAP mechanism, as deployed at RMA, pertained only to the use of the EDS itself; management of secondary waste was deferred to the RMA waste management plan (U.S. Army, 2001g). In accordance with that plan, neutralents were sent off-site to a commercial TSDF for incineration, and equipment rinsates and cleaning solutions were disposed of through the on-site wastewater treatment system.

Munitions Assessment and Processing System

The MAPS facility is currently under construction at Aberdeen Proving Ground (APG), Maryland. Construction and initial operation of the facility are being conducted under a RCRA RD&D permit. The current schedule is to have the facility completed by mid-2003, after which testing and operations will begin. The Army, in agreement with the state of Maryland, plans to transition from RD&D to a full RCRA operating permit once operations become routine. This progression will facilitate destruction of NSCWM stored at APG within the CWC schedule requirements. Thus far, the RAP process at APG has proceeded relatively smoothly, although there have been several construction delays.[5]

---

[1]William Brankowitz, Office of the PMNSCM, briefing to the committee on May 23, 2001.

[2]There has been much local criticism of the Army's handling of the Spring Valley incident. The concerns here seem to be focused on whether the Army should have performed a more thorough investigation of the area at an earlier date. The choice of RAP mechanism does not seem to have been a factor in the criticism.

[3]Department of Defense (DOD) facilities are required to consult with regulatory authorities and the public on decisions pertaining to removal actions, but the final decision belongs to DOD (CERCLA Section 212(f)). See Appendix F for details.

[4]Bill Brankowitz, Office of the PMNSCM, presentation to the committee on May 23, 2001.

[5]These delays were due to finding buried munitions-related debris at the MAPS construction site.

Pine Bluff Non-Stockpile Facility

The Army has just begun to evaluate permitting strategies for the PBNSF. As with MAPS, construction and initial operation of the facility could proceed under a RCRA RD&D permit. Then, after operations at the facility become routine, the PBNSF operations could be transitioned to the full RCRA permit.

SPECIFIC ISSUES

Regulatory Approval and Permitting Mechanisms

In the usual context, regulatory approval refers to the process of obtaining traditional RCRA permits from regulators for hazardous waste treatment, storage, and disposal facilities (TSDFs). However, there are more expeditious mechanisms under RCRA for obtaining the approval of the regulator, and several of these are well suited for the recovery and treatment of NSCWM, especially for small or even moderate finds or where mobile treatment technologies will be employed. In addition, in some cases, regulator approval for waste management activities may be obtained under CERCLA in lieu of RCRA. Accordingly, "RAP" refers to all mechanisms for obtaining regulator approval. RAP mechanisms are reviewed in Appendix F.

Cooperation Between the Army, the States, and the Public in the RAP Process

As indicated above, although there have been some bright spots (MAPS at APG and the EDS-1 at RMA) in the RAP process for the non-stockpile program, PMNSCM and some of the states have experienced excessive delays and unusually high expenses during the RAP process.

One of the state regulators' most frequent criticisms of the Army's non-stockpile permitting program is the Army's incomplete understanding of the regulatory requirements for permitting emerging technologies. Typically, the Army does not initially provide all the data and information that is required by regulators as input to the RAP process. Also, because non-stockpile technologies are innovative in many respects, the states do not always know ahead of time what data or information they need for complete RAP documentation. In addition, some state regulators perceive that the Army does not initiate the permitting process soon enough, especially given the tight time lines imposed by the CWC.

One of the Army's most frequent criticisms of state regulators is their tendency to be overly conservative in the regulation of chemical agents and associated wastes. Other common criticisms of state regulatory programs include their tendency to change and remake decisions regarding how the chemical agents should be regulated and, even after state regulations have been promulgated, to attempt to impose standards that are more stringent or broader in scope than those specified in the promulgated regulations.

One of the most frequent criticisms about the role of public stakeholders in the RAP process is that "public" opinion is very often driven by one or more public interest groups that do not adequately reflect the views of the local public. This criticism emanates from both the Army and state regulators. Public stakeholders, for their part, often perceive that the Army—and, to a lesser extent, the states—makes decisions about, for instance, which technology should be used for a particular application and then seeks public approval after the fact. Public involvement is discussed more completely in Chapter 5 of this report.

Classification of Chemical Agent Identification Sets

On June 26, 2001, the U.S. Army revised Army Regulation AR 50-6 for Chemical Surety. The surety program imposes a set of reliability, safety, and security control measures to ensure that only personnel who meet the highest standards of reliability conduct chemical agent operations, that chemical agent operations are conducted safely, and that chemical agents are secure. AR 50-6 applies to all Army activities involving chemical agents, including the stockpile and non-stockpile destruction programs.

One of the more significant revisions in AR 50-6 is the classification of CAIS as recovered chemical warfare materiel (RCWM). The committee believes, however, that it is not necessary to handle CAIS as RCWM. The volume of agent is small, there are no explosives or energetic materiel associated with CAIS, the DOT container requirements are adequate, and the risk is likely to be within the range typically accepted for commercial hazardous waste (NRC, 1996a). The added costs of treatment and disposal would be out of proportion to the benefits, if any, that would be realized by this decision. Furthermore, the committee believes that classification of CAIS as RCWM is unwarranted because the surety risks that this materiel poses in no way compare to the risks posed by other RCWM, such as recovered mortars or bombs (NRC, 1999a).[6]

Diverse Army Organizations with Responsibility for RAP

NSCMP is responsible for developing and proving technologies that are capable of destroying non-stockpile materiel and for treating secondary wastes and related waste in

---

[6]As discussed in NRC 1999a, this conclusion is based on the fact that CAIS sets contain no explosives and relatively small quantities of agent, and that the hazards of the chemical agents in CAIS (mustard and lewisite) fall within the range of the hazards presented by industrial chemicals ordinarily disposed of according to U.S. hazardous waste regulations.

compliance with regulatory requirements. However, individual installations are responsible for achieving RAP for the management of non-stockpile materiel at their installations. In cases where a developed technology is being demonstrated, such as through a RCRA RD&D permit, the NSCMP has an active role in achieving RAP. In general, however, it plays a secondary, technical support role in achieving RAP for NSCWM treatment at specific installations.

When non-stockpile materiel is found at nonmilitary locations, such as Spring Valley (Washington, D.C.), typically the U.S. Army Corps of Engineers is delegated responsibility for obtaining RAP for management of this materiel and implementing the cleanup program. Again, the NSCMP has a secondary role in such activities.

Other Army organizations also support the RAP process at specific locations, such as the Soldier and Biological Chemical Command (SBCCOM), the Army Environmental Center, and the Center for Health Promotion and Preventive Medicine. In addition, SBCCOM is responsible for permitted storage of NSCWM (and stockpile materiel) at installations where this materiel is stored.[7]

Overall, the missions of the entities responsible for supporting and achieving RAP for chemical materiel storage, stockpile destruction, and the non-stockpile program do not appear to be well coordinated within the Army. Also, the roles and responsibilities of the various Army entities involved in the RAP process for non-stockpile activities (and, in general, for all of the Army's chemical agent programs) and the interrelationships among them do not appear to be well defined. The diversity of Army entities involved in achieving RAP for chemical agent operations and the lack of coordination have caused confusion among the regulators and the public. State regulators and public interest groups have observed that the diverse Army entities involved in achieving RAP for chemical agent operations and deployment are often at odds with one another because they have competing missions and because communication among them is less than ideal.

Schedule Requirements of the CWC

The Army is bound by the requirements of the CWC in its destruction of stockpile and non-stockpile munitions. Although treaties ratified by Congress have the force and effect of law and do apply to the states, the CWC does not impose specific requirements directly on state regulators. In addition, although the states would prefer that chemical munitions be destroyed as quickly as possible, their primary concern is that the destruction occur in a manner that provides adequate protection to human health and the environment. There is no direct incentive for the states to meet CWC schedule requirements.

At the same time, the Army must receive regulatory approvals from the states to handle and treat non-stockpile materiel, so it must address the regulatory concerns of the states. As indicated previously, these concerns have sometimes resulted in very significant delays in the RAP process (e.g., for the MMD-1 and the RRS). Regulatory delays may make attainment of treaty deadlines difficult.

Part of the problem here is that states often have been faced with the challenge of evaluating and approving RAP documentation for stockpile and non-stockpile operations during the same time period. Another part is that many of the technologies being developed for NSCWM are innovative and have not previously been permitted in the United States. At the same time, the hazards posed by NSCWM present unusual concerns for the states and the public. Thus, the states are faced with reviewing innovative technologies and addressing special concerns posed by non-stockpile, often at the same time as they are processing permit documentation for stockpile operations.

In consideration of these circumstances, the committee believes that the states may not possess sufficient resources to make regulatory determinations for the non-stockpile program within the time frame required by the CWC schedule. While the Army provides funds to the states under cooperative agreements to oversee chemical agent operations, it is unclear whether the states are being provided sufficient funds given these unique challenges.

Overall Lack of a Regulatory Program for Treatment Requirements

Because there are no treatment standards for chemical agents and associated wastes established in federal or state regulations, treatment requirements are negotiated between the Army and the states on a case-specific basis. Further, the chemicals and technologies being regulated as part of the NSCWM have few commercial hazardous waste analogues. As a result, a large share of state resources must be devoted to developing regulations that affect a very low volume of waste, compared with commercial hazardous waste regulations. This has resulted in considerable delays in the RAP process at the state levels, often resulting in different treatment requirements being applied for the same type of chemical agent or secondary waste in different states.

Several states have considered developing regulations for chemical agent waste treatment. In 1995, the state of Utah and the Army began a cooperative effort to develop RCRA land disposal restrictions (LDR) for agent wastes. While the Army and the state met frequently on the rule and agreed to resolve most issues, several issues remained. Also, whereas

---

[7]These and other Army organizations may be involved in future land use planning, which is also an important consideration. However, future land use planning and related issues are beyond the scope of this report.

the initial concept was for the Army and the state to team up in writing the rule, the Army ended up writing its version and presenting it to the state for consideration.

In May 1999, the Army presented its version of the so-called Utah Chemical Agent Rule (UCAR) to the state of Utah for consideration (U.S. Army, 1999b). Three general principles were established, in agreement with Utah officials, as a basis for the rule:

1. Chemical agents should be regulated in the same manner as similar toxic materials generated by private industry in the state of Utah.

2. The primary basis for determining the level or stringency of regulation should be the potential risk that a release of a substance would pose to human health and the environment.

3. There should be a reasonably acceptable relationship between the cost of a regulation and its anticipated benefits.

In accordance with these principles, a risk-based approach was the underpinning for the rule. The Army's proposed UCAR addressed treatment technologies for chemical agents and associated wastes, including treatment technologies under development.[8] The rule was also expanded to address exemption levels—de minimis levels below which wastes that might contain (or have contained) chemical agents would no longer be considered hazardous waste.[9]

After the Army's proposed UCAR was presented to Utah officials, it was provided to other states that were interested in evaluating its provisions, primarily the stockpile states. Some states, including Utah, looked at the Army's proposed UCAR very carefully, but to date, no state has officially adopted it, or any parts of it.

The benefit of such a rule is that instead of defining and specifying treatment requirements on a case-by-case and state-by-state basis (as part of RAP documentation), requirements would be established in a clear, concise manner and would be applicable to a variety of cases and states. The committee believes that the UCAR represents a good starting point for further development of a rule that would establish treatment requirements for agent waste in a manner protective of human health and the environment and at the least possible cost. Further, the committee believes that significant savings in time and effort could be realized if such a rule were developed cooperatively among the primary states responsible for chemical agent and munitions treatment. The public and other stakeholders should be involved in any rule-making effort.

Secondary Waste Classification

Secondary wastes from agent destruction processes (such as occur in the EDS) include the primary reaction products, called neutralents, the excess reagent dilute aqueous rinses of the reaction vessel, and cleaning solutions used to remove residuals before processing the next NSCWM item or agent (NRC, 2001a). Secondary wastes will also include residuals from further treatment of neutralent, if further treatment is performed and such treatment generates additional treatment residuals.

Three issues associated with the regulatory classification of secondary wastes warrant discussion. The first is whether secondary wastes are acutely hazardous, as defined under RCRA.[10]

There is no question that the primary chemical agents found in non-stockpile items are acutely hazardous and warrant stringent controls. Secondary treatment residuals, however, are often placed in the same acutely hazardous category as the parent agent (as, for example, in the Utah and Colorado regulatory programs). RCRA requirements for management of acutely hazardous wastes are more restrictive than those for other types of hazardous waste—for example, there are tighter restrictions on the amount that may be stored at any one time. At worst, neutralents may contain trace amounts of agent, small amounts of agent degradation products, and unreacted reagents. The data gathered to date demonstrate that neutralization destroys 99.995 percent to 99.99998 percent of various chemical agents, so it is highly unlikely that neutralent would pose an acute toxicity risk. In fact, the Army has performed toxicity testing on neutralents and similar wastes that supports the contention that these types of waste are not acutely toxic in accordance with the RCRA definition (NRC, 2001a). While the handling and disposal of neutralent does warrant regulation and further treatment as a hazardous waste, the data do not support the stringent controls associated with the acute toxicity classification (NRC, 2001a, pp. 34-36).

The second issue is whether neutralent waste streams, as indicated above, contain trace amounts of agents and small amounts of agent degradation products. In most states where agent wastes are regulated as listed hazardous waste, neutralents, as a direct by-product of agent neutralization, would remain in the chemical agent hazardous waste category. The Army has indicated that it will pursue a hazard-

---

[8]RAP documentation for the MMD-1 and RRS was under development as the UCAR was being developed.

[9]These exemption levels are similar in concept to those that EPA is establishing for listed hazardous waste under the Hazardous Waste Identification Rule (HWIR). HWIR is discussed in Appendix F.

[10]Acutely toxic wastes under RCRA are wastes having a rat oral LD50 (lethal dose to 50 percent of exposed population) of less than 50 mg/kg, a rat inhalation LC50 (concentration lethal to 50 percent of exposed population) of less than 2 mg/L, or a rabbit dermal LD50 of less than 200 mg/kg (40 CFR 261.11(a)(2)).

ous waste delisting for the mustard hydrolysate (similar to neutralent in nature and content) that will be generated at APG, which would have the effect of removing the agent hazardous waste classification (NRC, 2001d). The hydrolysate would most likely still be classified as hazardous waste, however, because it is likely to exhibit one or more RCRA hazardous waste characteristics (see Appendix F for information on these characteristics). The advantage, in this case, is that removal of the agent from the listing would make it easier for off-site commercial TSDFs to accept the wastes for further treatment. Delistings, as reviewed in Appendix F, are typically long and arduous processes where data are generated and presented to regulators in support of waste declassification. While a delisting effort may make sense in cases where large amounts of waste will be generated over a relatively short period of time, as with APG, they do not make sense for the non-stockpile program, where different types of neutralents are likely to be generated at multiple locations in multiple states over a period of perhaps many years. The Army could consider, however, working with the states to establish a de minimis concentration for the agents in waste streams, which could be incorporated into the listing regulations. In this manner, neutralents would no longer be associated with the parent agent waste, and acceptance by off-site commercial TSDFs would be facilitated.

Third, some secondary wastes may not warrant regulation as a hazardous waste at all. Rinsates and cleaning solutions consist primarily of water with much lower concentrations of hazardous chemicals than the initial neutralent and appear to pose relatively little risk (NRC, 2001c). Yet in some states, they would also be regulated as acutely hazardous wastes, just as the parent agent is. In addition, residuals from the further treatment of neutralent, such as through chemical oxidation, are expected to be similar to rinsates and cleaning solutions in terms of risk. Not only does this materiel not warrant regulation as an acutely hazardous waste, but the committee believes that it might be safely managed as nonhazardous waste. To ensure that secondary wastes are classified properly for regulatory purposes, the Army should work with regulators and the public.

RAP for Mobile Technologies

The Army developed the RRS and EDS as mobile treatment technologies. Thus, they may be brought to a site where NSCWM are discovered. Use of a mobile treatment system such as the RRS or EDS has two stages. The first stage involves the following.

- transporting the system to the site
- obtaining the regulatory approvals necessary to operate the mobile treatment system
- actual neutralization of non-stockpile chemical materiel

The second stage involves the subsequent treatment and disposal of the RRS or EDS secondary waste streams, whether at the site where they are generated or at an off-site facility. Although the RAP mechanism applied for operation of the RRS or EDS device itself will affect the choice of RAP mechanism for treatment of the RRS or EDS secondary waste streams, the two are nevertheless distinct.

As indicated in Appendix F, operation of the RRS or EDS devices may be permitted under a number of mechanisms, including RCRA emergency permits or CERCLA emergency removal actions. RCRA orders may also be used. In more urgent emergencies, prior regulatory approval is not required for the initial response (EPA, 1997).

A number of RAP mechanisms could be used for the management of liquid waste streams from the second stage of RRS or EDS operations.

In some cases, secondary waste treatment could be deployed to the site where primary treatment is conducted. Given potential public opposition to sending secondary wastes off-site, this may be the preferred option. However, as indicated previously, secondary wastes from mobile treatment technologies are not expected to contain amounts of chemical agent or other chemicals that would render the materiel acutely hazardous. The committee believes that this materiel may safely be transported off-site or off-installation to a TSDF for treatment. In this case, the treatment would be conducted under the permit for an off-site TSDF. If the permit for the off-site TSDF is not written broadly enough to allow treatment of these secondary wastes, it may need to be modified. In this case, provisions for permit modification should be considered well ahead of initial treatment of the NSCWM. See Appendix F for additional information on RAP alternatives.

Certain state regulators (e.g., regulators in Oregon) have been considering requiring that wastes be agent-free prior to their release to off-site facilities. Resolution of this issue has contributed to the delay in issuing regulatory approvals and permitting decisions for the stockpile program. Such an approach, if adopted by other states, is expected to be a serious problem for the non-stockpile program, because materiel could be found virtually anywhere, and in many cases, off-site transport and treatment of secondary waste may be one of the better options from a risk and cost standpoint. Further, in the non-stockpile program, it may be necessary to ship secondary wastes that contain very small amounts of agents (as from CAIS or single-round containers (SRCs) to off-site facilities for treatment).

While an agent-free level could be based on risk, the state of Oregon is pursuing a definition based on analytical detection capability. The committee believes, however, that it is neither necessary nor appropriate to establish a definition of agent-free on any basis. Although the committee believes very strongly in controlling munitions that contain or may contain significant amounts of chemical agents, neutralization, as indicated previously, results in the destruction of between 99.995 percent and 99.99998 percent of the agent. For this reason, the committee sees little benefit in requiring that secondary wastes be totally agent-free prior to their release to off-site facilities for further management, and it sees no risk basis for requiring treatment until residuals are agent-free. EPA, for instance, uses risk- or technology-based approaches in establishing regulatory levels of concern; no federal environmental regulations are premised on the total absence of risk.

FINDINGS AND RECOMMENDATIONS

Regulatory Permitting/Approval Mechanisms

Finding 4-1. While the traditional RCRA permit is suitable for facilities like MAPS and PBNSF, initial operation of these facilities may be expedited using other less arduous RAP mechanisms. In addition, when mobile treatment systems or technologies are employed, and particularly for small or even moderate NSCWM finds, operations may be expedited using other (non-RCRA permit) regulatory approval mechanisms under RCRA or CERCLA.

Recommendation 4-1. The Army should work with state regulators to tailor RAP mechanisms to the magnitude of the NSCWM recovery and treatment operations. For facilities, initial operations should be conducted under expedited RAP mechanisms (e.g., a Research, Development, and Demonstration permit); traditional Resource Conservation, and Recovery Act (RCRA) permits, if necessary, should be employed after operations become routine. When mobile treatment systems or technologies are employed, and particularly for small or even moderate quantities of newly discovered NSCWM, expedited (non-RCRA permit) regulatory approval mechanisms under RCRA or the Comprehensive Environmental Response, Compensation and Liability Act (CERCLA) should be used, as appropriate.

Cooperation Between the Army, the States, and the Public in the RAP Process

Finding 4-2. The PMNSCM and the states have experienced excessive delays and incurred significant expenses in the RAP process. All of the primary stakeholders—the Army, state regulators, and public interest groups—share responsibility for this situation. If RAP is to be achieved in a timely and efficient manner in the future, early interaction of all stakeholders in the technology selection and RAP process is essential.

Recommendation 4-2. The Army should establish a prepermitting process to resolve RAP issues involving the Army, regulators, and the public for both mobile systems and non-stockpile treatment facilities. In addition, the Army should develop guidance on RAP for management of NSCWM. A guidance that is jointly issued by the Army and regulators, with input from the public, should be considered, and the committee recommends that it be of national scope.

Classification of CAIS

Finding 4-3. The Army's revision of Army Regulation 50-6, in which CAIS is now classified as RCWM, effectively puts CAIS in the same hazard category as recovered bombs and mortar rounds. The committee believes that this classification is inappropriate and that CAIS could safely be classified as hazardous waste.

Recommendation 4-3. The Army should reverse its classification of CAIS as recovered chemical warfare materiel (RCWM), thus avoiding additional time and cost for their destruction.

Diverse Army Organizations with a Responsibility for RAP

Finding 4-4. Several Army entities are involved in developing and demonstrating technologies for destroying NSCWM and treating secondary wastes and for achieving RAP for these activities. In addition, the Army has established separate RAP responsibilities for chemical warfare materiel (CWM) storage and for stockpile destruction operations. Army entities involved in these RAP processes often have competing missions, and communication among them is less than ideal. Further, the roles and responsibilities of these entities, and the interrelationships among them, are not well defined, and this has caused confusion for regulators and the public.

Recommendation 4-4. RAP for all of the Army's chemical agent programs, including the non-stockpile program, should be seamless and transparent to the regulator and the public, who should "see" only one Army across all chemical agent programs at a specific location or operation. An installation-specific (or in the case of off-site NSCWM finds, operation-specific) core Army RAP team should be established for all chemical agent operations, including treatment of NSCWM. Installation or operation representatives should lead the RAP team at each location. The team should be directed by a central Army organization encompassing all chemical agent operations that require RAP so as to promote communication, continuity, and consistency among them. This organization should have the authority to establish RAP policy for all chemical agent operations nationwide.

REGULATORY APPROVAL AND PERMITTING ISSUES

Schedule Requirements of the CWC

Finding 4-5. Most of the alternative technologies being developed for NSCWM are innovative and have not been previously permitted in the United States. At the same time, the hazards posed by NSCWM present special concerns for the states and the public. The states have also been faced with the challenge of evaluating RAP documentation for non-stockpile treatment at the same time that similar documentation for the stockpile program is under review. While the states have been provided funds, through cooperative agreements, to oversee the Army's chemical agent operations, it is unclear, given these challenges, whether those funds have been sufficient for them to evaluate NSCWM technologies and process RAP documentation within a time frame consistent with CWC deadlines.

Recommendation 4-5. The Army should examine funding provided to the states as part of existing cooperative agreements to ensure that they are sufficient to evaluate new or innovative NSCWM treatment technologies within a time frame consistent with CWC deadlines.

Overall Lack of a Regulatory Program for Treatment Requirements

Finding 4-6. Although the Army and the state of Utah worked together to develop regulatory-based chemical agent treatment standards (as part of the Army's proposed UCAR) and other states have examined these standards, the Army and the states have not continued joint efforts toward adoption of these standards. As a result, there are still no regulations that establish treatment standards for these wastes.

Recommendation 4-6. The Army and the states should continue to work together to achieve mutually acceptable regulations that define appropriate treatment for chemical agents and associated wastes. While state-specific treatment standards can be established, the committee recommends standards that are national in scope.

Secondary Waste Classification

Finding 4-7a. De minimis secondary wastes such as neutralents, rinsates, and cleaning solutions are classified as "acutely hazardous" by some states because they are derived from chemical warfare agents. Such classification results in more stringent management requirements. While neutralents may contain very minute amounts of agent and agent degradation products, it is unlikely that they would pose an acute toxicity hazard.

Finding 4-7b. While neutralents may exhibit hazardous waste characteristics, they contain such small amounts of agent that they could be considered as no longer associated with the parent agent waste. If the hazardous waste listing is reversed, off-site commercial TSDFs would be able to more easily accept the waste for treatment.

Finding 4-7c. Rinsates and cleaning solutions, as well as residuals from treatment of neutralent, will consist primarily of water and pose even less of a hazard than neutralent.

Recommendation 4-7. In states where secondary waste streams are regulated as acutely hazardous, the Army should work with state regulators to remove the designation "acutely hazardous." For neutralents, the Army should work with state regulators to establish de minimis concentrations for the agents in waste streams, to be incorporated into the listing regulations, whereby the waste would longer be considered as being associated with the parent agent waste. Further, the Army and the states should consider whether rinsates and cleaning solutions and residuals from the treatment of neutralent should be classified as hazardous waste at all.

RAP for Mobile Technologies

Finding 4-8. Certain state regulators have been considering requiring that secondary wastes be agent-free before they are released to off-site facilities. This would pose a serious problem for the non-stockpile program, because NSCWM can be recovered anywhere and the off-site transport of secondary wastes may be a viable management option.

Recommendation 4-8. Given the similarities between NSCWM secondary wastes and industrial hazardous wastes, the committee recommends that no additional prohibitions be placed on the off-site transportation of secondary wastes.

5

Public Involvement

As demonstrated in the literature and by the Army's own experience in the chemical stockpile program, public involvement is key to the timely achievement of the Non-Stockpile Chemical Materiel Product's (NSCMP's) mission (NRC, 1994, 1996a, 1996b, 2000a, 2001a). The previous reports noted that the public should be thought of not as monolithic but as different "publics"— that is, stakeholders[1] whose interests, level of awareness and information, and desired level of involvement vary. Facilitating their input to a policy or technology and their understanding and ultimate acceptance of it involves identifying interested or affected stakeholders, providing open and timely information, discussing and clarifying the issues of concern, putting in place mechanisms to facilitate their engagement, and establishing procedures to evaluate the recommendations of these publics and to give them feedback on how and why their input was or was not used (NRC, 1996a, 1996c, 1999b, 2001a).

INFORMATION SOURCES

In evaluating public involvement issues facing the NSCMP and making recommendations on them for this report, the committee reviewed a schedule of NSCMP program activities; Army Regulation AR 360-1, which provides guidelines on the Army Public Affairs Program; and recent public involvement of the PMCD and its public outreach plans related to non-stockpile activities (Booz-Allen and Hamilton, 2000; U.S. Army, 2001h).

Committee members also monitored the Non-Stockpile Chemical Weapons Citizens' Coalition (NSCWCC) and the Chemical Weapons Working Group (CWWG) Web sites and other National Research Council (NRC) publications and activities, as well as environmental publications such as *Defense Environmental Alert* and *Superfund Report*. In addition, the committee reviewed formal and informal discussions with, and documents provided by, stakeholders during its earlier studies (NRC, 1999a, 2001a). Primary stakeholders with whom the committee had met previously included federal and state regulators, representatives from the NSCWCC, the Citizens' Advisory Technical Team (CATT) established under the ACWA Program, and the Core Group established by NSCMP.[2]

---

[1] The terms "public" and "stakeholder" are used interchangeably in the text to refer to the variety of individuals and groups that are interested in, may affect, or may be affected by NSCMP decisions. They include the Congress, which enacted the statutes requiring the Army to make a decision; local citizens who may be affected by the decision; national nonprofit groups involved in the public policy debate; contractors, who must implement decisions; and federal and state officials and regulatory agencies.

[2] The NSCWCC is a coalition of grass-roots organizations opposed to incineration. The ACWA/CATT is the four-member Citizens' Advisory Technical Team that was established by the ACWA program manager to work directly with the ACWA technical team and report back to the citizens' interest groups, as well as members of the entire dialogue established under the ACWA program to select and test technologies. The Core Group includes Army personnel from the chemical demilitarization program, representatives of regulatory agencies, and representatives of citizens' groups; it meets twice a year to exchange information about the non-stockpile program.

For this report, the committee sought the views of additional stakeholders on their program experience and recommendations to the Army concerning public involvement. The committee selected these stakeholders because their public involvement experience included non-stockpile issues (e.g., at Aberdeen Proving Ground) or focused solely on the non-stockpile program (Core Group and NSCWCC). The committee's plans to meet with members of the public at Pine Bluff Arsenal, where non-stockpile facilities are planned, were not realized because no community-wide, site-specific outreach and involvement activities had been initiated by the installation at the time of this writing. The committee conducted both formal and informal discussions with the following stakeholders:

- A subgroup of the committee met with chairpersons of the Restoration Advisory Board (RAB) and the stockpile Citizens' Advisory Commission, both at Aberdeen Proving Ground (APG), in March 2001.[3]
- One NRC committee member observed a RAB meeting at APG in July 2001, and a subgroup of the committee subsequently conducted a series of 1-hour telephone interviews with five RAB members representing a cross section of interests and backgrounds. Some of those interviewed were also members of the APG Superfund Citizens Coalition.
- A subcommittee conducted a 90-minute telephone interview with a representative of the NSCWCC.
- Two committee members observed a meeting of the Core Group in June 2001, gave a presentation on the committee's activities, and talked informally with Core Group members.
- Two committee members and the study director made an initial visit in August 2001 to Pine Bluff Arsenal and the White Hall Outreach Office, one of two offices established by PMCD. They met with outreach staff and technical staff associated with the stockpile and non-stockpile programs, as well as with the chairperson of the Pine Bluff Stockpile Citizens' Advisory Commission. Subsequently, two committee members observed the scoping meeting for the environmental impact statement (EIS), held in October 2001.

STAKEHOLDER VIEWS ON KEY PROGRAM ISSUES

In their discussions with the committee, stakeholders provided input on several key program issues. Although the groups that they represent cannot be considered statistically representative of the public at large, these stakeholders indicate the viewpoints of informed and active opinion leaders that the Army should consider as it develops its overall strategy and plans.[4]

There is almost universal opposition to importing out-of-state wastes that could result in a site becoming a "dumping ground." Mobile technologies are viewed as a way to address this concern—for example, a member of the APG/RAB said it is better for NSCMP to work on developing mobile technologies rather than on trying to find a state and a site willing to take everything. Many stakeholders acknowledged that there is no one answer for non-stockpile disposal: all situations are different and different options are needed. One person, for example, considered it a balancing problem and said that both permanent and mobile technologies are needed.

The use of stockpile facilities to dispose of non-stockpile items both within and between sites is generally viewed as a site-specific issue that depends on both technical feasibility and public acceptance. Public acceptance varies not only from site to site but also at individual sites. At APG, for example, some (but not all) citizens in Harford County with whom the committee spoke saw no problem, whereas citizens across the Chesapeake Bay in Kent County are strongly opposed. Some citizens viewed the use of stockpile facilities for non-stockpile items as a breach of trust and expressed concern that it could represent the first step in making their site a dumping ground for out-of-state waste (there seemed to be general agreement, however, that existing and planned facilities at APG will probably suffice for non-stockpile needs). At Pine Bluff, program staff informed committee members that state regulators are strongly opposed to use of the stockpile facility for non-stockpile materiel and indeed will not consider a permit modification to allow it.

Nonincineration technologies being developed by the NSCMP program appear to have broad acceptance. RAB members at APG strongly endorsed them as a needed ad-

---

[3]The mission of the RAB is to provide advice on cleanup to APG. Non-stockpile rounds, which may be found at APG, comprise a subset of operations. The board, which has been in operation for approximately 9 years, includes up to 20 members, selected by citizen members and appointed by the garrison commander. Several RAB members are, or have been, members of the Superfund Citizens' Commission and reported that the groups work together closely. The latter group's Technical Assistance Grant consultant (an independent consultant for citizen groups funded under a grant by the EPA) also plays an active role on the board. The mission of the Citizens' Advisory Commission (CAC) is to give the governor local citizens' input on stockpile issues; its charter has not been formally amended to include non-stockpile issues. The CAC was established in 1994; its members are appointed by the governor of Maryland.

[4]See, for example, the recommendations to focus on the views of the opinion leaders who "make things happen," provided by the director of the Massachusetts Military Reservation in "Lessons Learned from Environmental Mistakes," as reported in *Defense Environmental Alert,* August 28, 2001, p. 11.

vance over open detonation, which had the potential for emissions to the atmosphere as well as noise. One of the members said that the non-stockpile program has a good reputation and characterized the technologies as "decent first steps." This person expressed the view that the technologies are acceptable from an environmental perspective, both nationally and in the field. Members of the NSCWCC praised the Army for its efforts to develop nonincineration technologies that are both feasible and acceptable to the public.

Stakeholders generally had favorable views on deployment of the MAPS at APG and on the EDS. While MAPS was acknowledged as not providing a total solution since it does not destroy the agent, it was nevertheless supported by those with whom the committee spoke because it could deal with other wastes of community concern at APG as well as help to keep the Army from "using up" the EDS. However, as evidenced by a letter from a representative of the APG Superfund Citizens' Coalition, some members of the local community are concerned about the cost of MAPS and the associated potential for DOD to consider its use for out-of-state waste in order to justify that cost.[5] The EDS, in turn, was favored as a solution for materiel that must be destroyed immediately and would otherwise be detonated. One RAB member at APG, commenting on reported criticisms that the EDS is too costly and produces too much secondary waste, said that the first step in a new technology is never the cheapest and most efficient and that it has to be seen as something that will become a valuable tool as future generations are developed. Some Core Group members reported reservations about the Donovan blast chamber. While recognizing that it is able to handle more waste, they expressed concern that its effectiveness in destroying chemical agent had not yet been tested. Some also said that in judging a disposal technology, containment is more critical than destruction efficiency.

Some environmental activist groups continue to strongly oppose incineration—indeed, incineration is a hot-button issue that extends beyond the chemical weapons policy area. The NSCWCC continues to advocate storage of neutralent pending the development of nonincineration secondary technologies. It believes that postponing deployment of nonincineration alternatives is a viable option, given the slippage in meeting original CWC deadlines. An NSCWCC spokesperson said that broad public acceptance of plasma arc technology, currently under consideration for one of the facilities to be deployed at Pine Bluff, cannot be assured and hoped that NSCMP would not rush to put the technology in place. The spokesperson characterized plasma arc as a "synonym for incineration," with the associated disadvantages of high temperature and potential for formation of dioxins and reported that environmental and public health organizations around the world are very concerned about the extent to which the technology is being promoted by some vendors. Similar reluctance to embrace the technology has been expressed by a Core Group member and author of a recent report (Lynch, 2002).

As reported previously (NRC, 2001a), NSCWCC has recommended several criteria for technology selection. These criteria are consistent with the findings from earlier ACWA and AltTech studies (NRC, 1999b, 1996c):

• ability to contain by-products and effluents for analysis and reprocessing
• ability to identify by-products and effluents
• low-temperature and low-pressure operation
• no production of dioxins or furans
• incorporation of pollution prevention (i.e., as little generation of secondary waste as possible) (NRC, 2001a)[6]

The NSCWCC's advocacy of pollution prevention was also endorsed by a member of the APG RAB, although he noted that the cradle-to-grave concept of looking at a technology as a whole—taking into account the amount and type of secondary wastes produced—is moderately new to DOD. He added that this is a hard concept to get across to installations in the field that are faced with the everyday task of doing the work when there are too many tasks and too little time and money, and that it will take time for people in the defense community to view the issue holistically.

STAKEHOLDER VIEWS ON PUBLIC INVOLVEMENT

Everyone interviewed agreed on the need for public involvement in NSCMP decisions. When asked for suggestions on how the Army should deal with the communities near which it is proposing to construct new facilities,

---

[5]Letter from Cal Baier-Anderson, University of Maryland, Program in Toxicology, to James L. Bacon, PMCD, November 30, 2000.

[6]Providing information on technical issues such as these before selecting a treatment technology would facilitate dialogue with members of the public. Issues that have been raised frequently by citizen groups include the concentration of agent and hazardous constituents contained in the residuals from neutralization and other treatment technologies and the risks from treating, handling, and disposing of those residuals; emissions and effluents from the technology, and whether they can be held and tested prior to release; whether the technology is low temperature and low pressure; whether the technology minimizes the generation of secondary wastes; the relative advantages and disadvantages of the CWWG's recommendation that all non-stockpile materiel be neutralized first; and if virtually all agent is neutralized, the comparative advantages and disadvantages of storage, incineration of the neutralent, and alternative treatment.

interviewees cited three components of an effective approach and offered several recommendations:[7]

1. Early provision of information to the public—"openness," "transparency," "provision of as much information as possible," "deal proactively with the public," were frequent recommendations. RAB members at APG and the NSCWCC representative all commended the Army for its improved information flow and effort to be open with the public.

2. Willingness to go beyond the provision of information by listening to the public and establishing working relationships with it was recommended. At Aberdeen, for example, several members of the RAB commented favorably on the open provision of information, respect for citizens' views, and openness to incorporating these views and priorities into decisions. Most RAB members interviewed by the committee pointed to the effective working relationship between the Army and community that had resulted from such an approach.

3. Inclusion of the public in identifying and addressing the trade-offs inherent in developing strategies, systems, and facilities was recommended. All persons with whom the committee spoke acknowledged the complexity of the non-stockpile program and the challenges faced by management. Several recommended that the Army include the citizens who are most affected by Army decisions in the process of identifying and resolving trade-offs in the selection of disposal technologies and that it put in place mechanisms that allow active community input early in the process, when critical decisions are being made.[8]

For example, in discussions with the committee in open session, citizen representatives cited the dialogue process, including its linkage to affected communities, established by the ACWA program as a model for early and direct public involvement in technology decisions (NRC, 2001a).[9] In particular, they noted the trust engendered among all participants by a process in which public input was valued and recognized as an essential part of developing a solution. An APG member added that, regardless of the specific technology, there are critical considerations such as waste streams, long-term risks, and stability and that educated community members may have perspectives on these criteria that are very different from those of a military entity, regulatory agency, or private company. He expressed the view that to not consider these differing perspectives is foolish in this day and age.

It was apparent from the discussions that almost all of those interviewed rated the Army and the non-stockpile program highly for their performance on the first two public affairs components—specifically their increased openness in providing information and in listening to, and developing relationships with, a range of stakeholders. Some stakeholders said, however, that there is room for improvement, particularly in terms of providing *involvement* mechanisms for citizens to provide input to decisions. At APG, for example, one RAB member reported that, in his opinion, there has been "reasonable" involvement and that the program is headed in the right direction. However, he believes the non-stockpile program and the military in general have a long way to go in terms of understanding what adequate and effective public involvement really is.

In addition, some stakeholders said that NSCMP needs to recognize the public's ability to help NSCMP management achieve its program mission. One stakeholder said that many of the personnel are engineers who see the public as a stumbling block rather than a positive force that can work effectively with them to achieve their mission. This individual further noted that there is a lot of historical baggage, there needs to be greater trust in the public and regulators to come up with innovative ideas, and that the Army needs to make some shifts in its assumptions, based on success stories, to date. Another stakeholder said that the government should

---

[7] These components are generally consistent with the threefold division of public affairs activities provided in a letter report from the Committee on Review and Evaluation of the Army Chemical Stockpile Disposal Program. The components were public relations (provision of written information materials), outreach (opening channels of communication to the public so that its values, concerns, and needs can be heard), and involvement (development of a formal process that gives stakeholders the opportunity for input to decisions without surrendering the agency's legal mandate to make those decisions) (NRC, 2000a).

[8] A wide variety of mechanisms are available and have been discussed in the public involvement literature. These include citizen advisory boards and task forces, workshops, roundtables, dialogues, and brainstorming meetings. All of these mechanisms are designed to promote opportunities for the public to contribute to project decisions before a policy is set. In general, a mechanism will provide for representation of diverse stakeholders, open dialogue between project staff and the public, and an agency commitment to incorporating public input. For a practical discussion of particular mechanisms see Creighton (1985).

[9] The ACWA program initiated a public involvement group called the Dialogue, facilitated by the Keystone Center, consisting of 35 representatives from each state with a stockpile depot, members of such national groups as the Sierra Club, and the ACWA program manager. A key feature of the ACWA Dialogue was its focus on dialogue, consensus building, and problem solving, with the specific requirement by Congress to identify and demonstrate no fewer than two alternatives to the baseline incineration process: the public was involved in (i.e., actively engaged in) establishing criteria as well as in making trade-offs. Although PMNSCM has also organized a group of stakeholders (the Core Group), also facilitated by Keystone, the focus of the group is the exchange of information and opinions rather than dialogue, problem solving, and implementation of recommendations—i.e., there is a greater focus on outreach than on involvement and active engagement.

stop spending millions of dollars on a product that it only assumes will be acceptable to the community. He cited as an example the Army's preference for incineration, which it views as an efficient, working technology. He pointed out that if Army personnel had a relationship with people at a site, they would soon find out that incineration is not acceptable to the local community. He said he viewed involvement as a process of thinking ahead before developing plans and spending millions of dollars, but noted that this view goes against the institutional way of thinking. He believed strongly that the non-stockpile program needs to expand its involvement activities in order to move ahead.

Several success stories were related. A member of the RAB at APG noted how the RAB enhances the Army's ability to overcome its narrow technical focus and see the big picture from the community's perspective. Other members of the RAB cited specific ways in which RAB recommendations on facility design and emergency preparedness had been adopted by the Army. Some pointed to the value of public input in moving the Army from its previous focus on the most contaminated areas at APG to a new focus on land and waterway boundary areas, where the risk of public exposure is greatest. Core Group and NSCWCC representatives pointed to the valuable role that members of the informed public had played in facilitating EDS deployment at Rocky Mountain Arsenal. NSCMP acknowledged in Core Group meetings that the Core Group and other members of the public, who had been informed about the EDS by public affairs staff, played a very positive role by advocating the use of EDS in discussions with the EPA, state regulators, local officials, and skeptical members of the public in communities near the arsenal.

NSCMP PLANNING FOR PUBLIC INVOLVEMENT

Based on a review of program documents, several observations can be made about NSCMP public affairs activities. For this section the committee reviewed Army Regulation (AR) 360-1,[10] recent NSCMP public involvement planning documents (Booz-Allen and Hamilton, 2000; U.S. Army, 2001h), information provided at Core Group meetings, the schedule of NSCMP program activities, and information provided at the EIS scoping meeting in Pine Bluff and during the earlier visit to the arsenal of some members of the committee.

First (consistent with the information provided in the committee's discussions with stakeholders), activities recommended in AR 360-1 and described in the public affairs strategy and the Pine Bluff public outreach plan (Booz-Allen and Hamilton, 2000; U.S. Army, 2001h) focus almost exclusively on providing information and opening channels of communication to hear stakeholders' values, concerns, and needs (i.e., the first two components of a public affairs program).[11] While NSCMP is to be commended for its commitment to improving these activities, including placing a priority on interaction with Native Americans and on addressing environmental justice issues, Army and NSCMP management commitment to the third component, public involvement, is lacking. The committee's observation indicates that current activities—particularly at APG and with the Core Group—appear to be leading to improved dialogue and relationship building with a diversity of groups. However, the program does not include a means of linking national-level and local-community dialogues, either in the Core Group structure or in a formal process for influencing decisions, such as the ACWA dialogue, that is recognized and supported by senior Army decision makers. As demonstrated in the present report, there are many trade-offs to be addressed by the Army in developing strategies, systems, and facilities, yet the affected public is not formally involved in assessing these trade-offs.

Second, the project schedules provided to the committee by NSCMP reveal several weaknesses, particularly in relation to public involvement and permitting. In general, there is an inconsistent use of milestones, making it difficult to measure progress and perhaps leading to the neglect or delay of important issues. Several specific weaknesses, discussed below, appear to indicate that the involvement of stakeholders is viewed as a requirement of the law and not as an integral and valued part of NSCMP's overall decision-making process.

Specific weaknesses associated with public involvement include the following:

• Public involvement and permitting are treated as independent activities that have no effect on other activities or decisions, indicating the absence of a relationship between public contributions and program decisions and the apparently low value placed on the former.

---

[10]Army Regulation 360-1, dated September 15, 2000, states as follows: "This new regulation is a consolidation of several regulations that provide guidelines for command and public information, including information released to the media, and community relations programs intended for internal and external audiences with interest in the U.S. Army."

[11]The Pine Bluff Arsenal document (U.S. Army, 2001h), as indicated in its title, is limited to information and outreach activities. NSCMP's Mission and Vision Statement (Booz-Allen and Hamilton, 2000, p. 7) states as follows: "The mission of PMCD PIO is to provide a public information and involvement program that supports meaningful public participation and dialogue." However, no examples are provided of formal mechanisms whereby the public can be meaningfully involved in (i.e., engaged in) and able to provide input to the decision process.

- Only the minimum time called for by law is set aside for considering and incorporating public comments,[12] indicating an unrealistic time frame for public review by all (not only local) stakeholders, particularly in view of the stockpile program's experience of public opposition and associated program delays.
- Public involvement activities (other than time scheduled for public review of the National Environmental Policy Act and permitting documents) are linked only to the pre-operational and operational phases rather than to the construction and design phases of the non-stockpile facility proposed for Pine Bluff, a sign that "involvement" comes only after critical technical decisions have been made.
- Public involvement activities are included only for local stakeholders at Pine Bluff, indicating a gap between national and local activities and the potential for underestimating the impact of national stakeholders on program decisions.
- There is no mention of any of the alternative non-incineration technologies currently being designed and tested by NSCMP, indicating that they are not being considered for possible introduction.
- There is only limited training of managers in interacting with the public. The project schedules show one day of training in risk communication, with no indication as to the intended recipients (is the course for public affairs staff only, or for management?). Non-stockpile public affairs staff are to be commended, however, for having very recently provided awareness training on Native American perspectives for project managers. Similar training on African-American perspectives is planned in the future. However, no management training in interacting with other segments of the public is planned or referenced in the project schedules.

Third, in discussions with this committee, NSCMP spokespersons consistently emphasized the importance to the program of the plans being developed for a non-stockpile facility at Pine Bluff Arsenal. Yet, the limited plans for public involvement in Pine Bluff facility decisions contrasted sharply with the high level of RAB members' awareness of non-stockpile activities and the public involvement activities conducted by the garrison commander at Aberdeen Proving Ground. For example, as committtee members who visited the arsenal and the outreach office at Pine Bluff learned, community outreach and provision of site-specific information on the envisaged non-stockpile facilities and plans were not conducted among the Pine Bluff community until immediately before and during the EIS scoping meeting. PBA staff said that one reason for this lack of site-specific non-stockpile information and outreach was the higher priority assigned to stockpile program needs. Site personnel also believed that their local public is very supportive of arsenal activities and disinclined to become actively involved. Further research is needed to systematically characterize community views and interpret the limited public attendance at the scoping meeting and the paucity of public comments submitted on the Pine Bluff EIS. Experience at other stockpile sites shows the need for coordinated, consistent program-wide planning for and implementation of public involvement. In particular (1) continued local public support cannot be guaranteed and must be carefully nurtured, as it was at APG, and (2) attention must be paid to the full range of stakeholders likely to become active and have the ability to affect program decisions.

Finally, as evidenced in recent public involvement and public outreach plans for non-stockpile activities (Booz-Allen and Hamilton, 2000), NSCMP does not have the authority to direct information, outreach, and involvement programs at the many sites where non-stockpile chemical materiel may be found. The document specifies that authority resides with the installation commander who "will request assistance from NSCMP as appropriate" (Booz-Allen and Hamilton, 2000, p. 18). Thus, without agreement between NSCMP and the installation on program priorities, NSCMP is limited in its ability to develop public involvement activities that are consistent across sites, that respect national program priorities and needs, and that respond to local concerns and audiences.

FINDINGS AND RECOMMENDATIONS

Finding 5-1. NSCMP has become more open in providing information and in developing relationships with a range of stakeholders concerning NSCMP issues. The committee finds, however, that there is room for improvement in three areas:

- coordination of the missions of the program and the installations and associated public involvement plans and activities to ensure that NSCMP develops public involvement activities that are consistent across sites and concur with national priorities and needs
- expansion of public affairs programs to emphasize the development of involvement (as opposed to outreach) activities that engage the public at both the local and national levels and allow them to contribute to program decisions
- integration of local, site-specific public involvement activities and national-level public involvement activities into NSCMP project schedules to provide opportunities for all interested citizens to affect key decisions

Recommendation 5-1. As with RAP activities, public involvement should appear seamless across Army programs and transparent to local and national stakeholders. The committee recommends that the Army establish central direction

---

[12]In one instance, publication of the record of decision is shown as occurring on the same day that the public comment period ends.

to ensure coordination of program and installation missions and to promote continuity and consistency in public involvement programs across installations and between program and installation staff.

Recommendation 5-2. The committee recommends that the Army expand its public affairs program to include involvement as well as outreach activities. Specifically, for the Army to gain from lessons documented in studies of the stockpile program, the committee recommends as follows:

- The Army should direct installations to implement, in coordination with program staff, a strategy that includes development of public involvement mechanisms. Such mechanisms must be fully integrated with project schedules so that the public has a genuine opportunity to provide input to project decisions. Their goal must be to engage both the local public and other stakeholders in discussing and evaluating the various technologies being considered and to provide a continuing means of involving them in future planning efforts and project decisions.

- The Army should conduct public involvement training for program and installation personnel, including commanders, public relations, and program technical staff. Such training must be more extensive than a one-day training course in risk communication and must be conducted very early in the program. The training should be provided on a continuing basis to ensure adequate preparation of newly assigned personnel.

- NSCMP should consider how the program could more effectively use existing mechanisms, such as the Core Group, to include and engage citizens at the local level as well as the national level in identifying specific concerns and considering the trade-offs inherent in program decisions.

References

Arkansas. 1999. Permit for a Hazardous Waste Management Facility. Permit number 29-H. Little Rock, Ark.: Arkansas Department of Pollution Control and Ecology.

Blackwood, Milton E., Jr. 1998. Beyond the Chemical Weapons Stockpile: The Challenge of Non-stockpile Materiel. *Arms Control Today* 28(5). Also available at <http://www.armscontrol.org/act/1998_06-07/blajj98.asp>.

Booz-Allen and Hamilton, Inc. 2000. Public Involvement Strategy for the Non-Stockpile Chemical Materiel Project. March. Aberdeen Proving Ground, Md.: Program Manager for Chemical Demilitarization.

Burns and Roe. 2001. PLASMOX Technology for the Destruction of Chemical Warfare Materiel. Burns and Roe and MGC-Plasma AG. October. Oradell, N.J.: Burns and Roe Enterprises, Inc.

Creighton, J.L. 1985. *Public Involvement Guide*. U.S. Department of Energy, Bonneville Power Administration. Portland, Ore.

DeMil International. 2001. Donovan Blast Chamber. Available at <http://www.demil.net/blastchamber.htm>.

EPA (Environmental Protection Agency). 1997. Military Munitions Rule. 62 FR 6621, February 12. Washington, D.C.: Environmental Protection Agency.

EPA. 1998. On-Site Incineration: Overview of Superfund Operating Experience. EPA-542-R-97-012. March. Available at <www.epa.gov/swertio1/products/costperf/incinrtn/intro.htm>.

EPA. 1999a. NESHAP: Final Standards for Hazardous Air Pollutants for Hazardous Waste Combustors (Phase I) Rule. 64 FR 52828, September 30. Available at <www.epa.gov/epaoswer/hazwaste/combust>.

EPA. 1999b. Hazardous Waste Combusters, Frequently Asked Questions. Available at <www.epa.gov/epaoswer/hazwaste/combust/faqs.htm#dnf>.

EPA. 2000. Frequently Asked Questions on Final Rule on Hazardous Waste Combustion Emission Standards. Available at <www.epa.gov/epaoswer/hazwaste/combust/ faqs.htm> (last modified on 7/19/2000).

EPA. 2001a. NESHAP: Standards for Hazardous Air Pollutants for Hazardous Waste Combustors. 66 FR 35,087, July 3. Washington, D.C.: Environmental Protection Agency.

EPA. 2001b. Risk Burn Guidance for Hazardous Waste Combustion Facilities. EPA 530-R-01-001, July. Washington, D.C.: Environmental Protection Agency.

Flamm, K.J., Q. Kwan, and W.B. McNulty. 1987. Chemical Agent and Munition Disposal: Summary of the U.S. Army's Experience. Report SAPEO-CDE-IS-87005. Aberdeen Proving Ground, Md.: Program Manager for Chemical Demilitarization.

Greenpeace. 2001. Ban the Burn. Available at <www.greenpeace.org/~toxics>.

Lynch, P.E. 2002. Evaluation of PLASMA ARC Technology for Treatment of Non-Stockpile Chemical Warfare Materiel. February. Alameda, Calif.: Clearwater Revival Company.

NRC (National Research Council). 1994. Recommendations for the Disposal of Chemical Agents and Munitions. Committee on Review and Evaluation of the Army Chemical Stockpile Disposal Program. Washington, D.C.: National Academy Press.

NRC. 1996a. Public Involvement and the Army Chemical Stockpile Disposal Program, Letter Report from the Committee on Review and Evaluation of the Army Chemical Stockpile Disposal Program. Washington, D.C.: National Academy Press.

NRC. 1996b. Understanding Risk: Informing Decisions in a Democratic Society. Committee on Risk Characterization, National Research Council. Washington, D.C.: National Academy Press.

NRC. 1996c. Review and Evaluation of Alternative Chemical Disposal Technologies. Panel on Review and Evaluation of Alternative Chemical Disposal Technologies, Board on Army Sciences and Technology. Washington, D.C.: National Academy Press.

NRC. 1998. Using Supercritical Water Oxidation to Treat Hydrolysate from VX Neutralization. Committee on Review and Evaluation of the Army Chemical Stockpile Disposal Program. Washington, D.C.: National Academy Press.

NRC. 1999a. Disposal of Chemical Agent Identification Sets. Committee on Review and Evaluation of the Army Non-Stockpile Chemical Materiel Disposal Program. Washington, D.C.: National Academy Press.

NRC. 1999b. Review and Evaluation of Alternative Technologies for Demilitarization of Assembled Chemical Weapons. Committee on Review and Evaluation of Alternative Technologies for Demilitarization of Assembled Chemical Weapons. Washington, D.C.: National Academy Press.

NRC. 2000a. Evaluation of Demonstration Test Results of Alternative Technologies for Demilitarization of Assembled Chemical Weapons: A

Supplemental Review. Committee on Review and Evaluation of Alternative Technologies for Demilitarization of Assembled Chemical Weapons. Washington, D.C.: National Academy Press.

NRC 2000b. A Review of the Army's Public Affairs Efforts in Support of the Chemical Stockpile Disposal Program. Letter Report from the Committee on Review and Evaluation of the Army Chemical Stockpile Disposal Program. Washington, D.C.: National Academy Press.

NRC. 2001a. Evaluation of Alternative Technologies for Disposal of Liquid Wastes from the Explosive Destruction System. Committee on Review and Evaluation of the Army Non-Stockpile Chemical Materiel Disposal Program. Washington, D.C.: National Academy Press.

NRC. 2001b. Analysis of Engineering Design Studies for Demilitarization of Assembled Chemical Weapons at Pueblo Chemical Depot. Committee on Review and Evaluation of Alternative Technologies for Demilitarization of Assembled Chemical Weapons. Washington, D.C.: National Academy Press.

NRC. 2001c. Disposal of Neutralent Wastes. Committee on Review and Evaluation of the Army Non-Stockpile Chemical Materiel Disposal Program. Washington, D.C.: National Academy Press.

NRC. 2001d. Review of Proposed Process Changes for Expedited Disposal of the Aberdeen Stockpile of Bulk Mustard Agent. Ad Hoc Panel on Review of Proposed Process Changes for Expedited Disposal of the Aberdeen Stockpile of Bulk Mustard Agent. Washington, D.C.: National Academy Press.

NRC. 2002a. Closure of the Johnston Atoll Chemical Disposal System. Committee on Review and Evaluation of the Army Chemical Stockpile Disposal Program. Washington, D.C.: National Academy Press.

NRC. 2002b. Update on the Engineering Design Studies evaluated in the NRC report "Analysis of Engineering Design Studies for Demilitarization of Assembled Chemical Weapons at Pueblo Chemical Depot." Letter Report from the Committee on Review and Evaluation of Alternative Technologies for Demilitarization of Assembled Chemical Weapons. Washington, D.C.: National Academy Press.

OPCW (Organization for the Prohibition of Chemical Weapons). 2001. Unclassified Technical Assistance Visit Final Report. August 14. The Hague, Netherlands: Organization for the Prohibition of Chemical Weapons.

Oregon. 1997. Permit for the Storage and Treatment of Hazardous Waste, Umatilla Chemical Agent Disposal Facility. ID No. ORQ 000 009 431. Portland, Ore.: Department of Environmental Quality.

Sandia National Laboratories. 2000. Explosive Destruction System Fatigue/Life Cycle Analysis. Albuquerque: Sandia National Laboratories.

Sierra Club. 2001. Guidance On Hazardous Waste Incinerators. Available at <www.sierraclub.org/toxics/factsheets/incineration.asp>.

Smithson, A.E. 1994. The U.S. Chemical Weapons Destruction Program: Views, Analysis, and Recommendations. Report Number 13. September 1994. Washington, D.C.: The Henry L. Stimson Center.

Stone & Webster. 2001a. Evaluation of Donovan Blast Chamber for Destruction of Chemical Warfare Materiel, T-130. Prepared by Stone & Webster, Inc., for Program Manager for Chemical Demilitarization. February 15. Aberdeen Proving Ground, Md.: Program Manager for Chemical Demilitarization.

Stone & Webster. 2001b. Evaluation of the Limited Engineering-Scale Testing of the PLASMOX® Technology to Treat Chemical Warfare Materiel. July. Aberdeen Proving Ground, Md.: Program Manager for Chemical Demilitarization.

U.S. Army. 1993. Survey and Analysis Report. Prepared by the Project Manager for Non-Stockpile Chemical Materiel. Aberdeen Proving Ground, Md.: Program Manager for Chemical Demilitarization.

U.S. Army. 1996. Survey and Analysis Report, 2nd edition (draft). Prepared by the Project Manager for Non-Stockpile Chemical Materiel. December. Aberdeen Proving Ground, Md.: U.S. Army Program Manager for Chemical Demilitarization.

U.S. Army. 1999a. Binary Chemical Weapons. NSCMP information products brochure. Available at <http://www.pmcd.apgea.army.mil:80/nscmp/ip/bro/nacmp/binary.asp>.

U.S. Army. 1999b. U.S. Army Proposed Utah Chemical Agent Rule (UCAR) for Consideration by the State of Utah, Volumes 1 and 2, U.S. Army Land Disposal Restrictions–Utah Group (LDRUG). May. Aberdeen Proving Ground, Md.: Program Manager for Chemical Demilitarization.

U.S. Army. 2000a. Chemical Weapons Destruction Complete on Johnston Atoll. Office of the Assistant Secretary of Defense, Public Affairs. November 30. Washington, D.C.

U.S. Army. 2000b. Test of the Drill Through-Valve System at Porton Down, United Kingdom (Phase 2: Live Agent Tests), Final (Version 2). November. Aberdeen Proving Ground, Md.: Program Manager for Chemical Demilitarization.

U.S. Army. 2001a. U.S. Army Chemical Demilitarization Program Releases. Updated Official Schedule and Cost Estimates, Press Release. Information Office, October 4. Aberdeen Proving Ground, Md.: Program Manager for Chemical Demilitarization.

U.S. Army. 2001b. PMNSCM Schedule for Treatment of Non-Stockpile Inventory. September 25. Aberdeen Proving Ground, Md.: Program Manager for Chemical Demilitarization.

U.S. Army. 2001c. Transportable Treatment Systems for Non-Stockpile Chemical Warfare Materiel. Final Programmatic Environmental Statement. Volume 1. Aberdeen Proving Ground, Md.: Program Manager for Chemical Demilitarization.

U.S. Army. 2001d. Non-Stockpile Systems Program, Rapid Response System (RRS) After Action Report. Contract No. DAAA-09-95-D-0001, Task 0019. November. Prepared by Teledyne Brown Engineering.

U.S. Army. 2001e. Rapid Response System Test Report. Final. October. Aberdeen Proving Ground, Md.: Program Manager for Chemical Demilitarization.

U.S. Army. 2001f. Memorandum for Record: Initial Assessment of the Donovan Controlled Chamber (CDC) Used in Belgium from 14 May through 22 June. Charles L. Twing, U.S. Army Corps of Engineers, Huntsville Engineering and Support Center, June 28. Aberdeen Proving Ground, Md.: Program Manager for Chemical Demilitarization.

U.S. Army. 2001g. Boneyard Clearance Project. Waste Management Plan. Appendix T, Section 36. Prepared for Rocky Mountain Remediation Venture Office. Morris Plains, N.J.: Foster Wheeler Environmental Corporation.

U.S. Army. 2001h. Non-Stockpile Materiel Project Outreach Plan for Pine Bluff Arsenal, Final Working Draft. March. Aberdeen Proving Ground, Md.: Program Manager for Chemical Demilitarization.

U.S. Army. 2002a. Spring Valley, Washington, D.C., Project Overview, U.S. Army Corps of Engineers Baltimore District. Available at <http://www.nab.usace.army.mil/projects/WashingtonDC/springvalley/overview.htm>.

U.S. Army. 2002b. Formerly Used Defense Sites, U.S. Army Corps of Engineers. Available at <http://hq.environmental.usace.army.mil/programs/fuds/fuds.html>.

Yang, Y. 1995. Chemical Reactions for Neutralizing Chemical Warfare Agents. *Chemistry and Industry* 8(May 1): 334-337.

Zepata Engineering. 2000. Tests of Blast Chamber for Destruction of Chemical Warfare Materiel. Draft 1 Engineering Report. September. Huntsville, Ala.: U.S. Army Engineering and Support Center.

Appendixes

A

Biographical Sketches of Committee Members

John B. Carberry, chair, is director of environmental technology for E.I. duPont de Nemours and Company, where he has been employed since 1965. He is responsible for providing technical analysis of existing and emerging environmental issues. Since 1988, he has been involved with initiatives to advance DuPont's environmental excellence through changes in products, recycling of materials, and renewal of processes, with an emphasis on reducing waste and promoting affordable, publicly acceptable technologies for the abatement, treatment, and remediation of environmental pollution. Mr. Carberry is chairman of the Chemical Engineering Advisory Board at Cornell University, a fellow of the American Institute of Chemical Engineers, and a member of the Radioactive Waste Retrieval Technology Review Group for the U.S. Department of Energy. He was a member of the National Academy of Engineering (NAE) Committee on Industrial Environmental Performance Metrics. He holds an M.S. in chemical engineering from Cornell University and an M.B.A. from the University of Delaware.

John C. Allen is vice president of transportation at Battelle Memorial Institute. Prior to joining Battelle in 1983, he served as transportation economist and policy analyst with the U.S. Department of Transportation's Office of Hazardous Materiel Transportation. Mr. Allen has managed and participated in numerous studies involving the policy, regulatory, institutional, and safety aspects of transporting hazardous and nuclear materials. He has served on various National Research Council (NRC) advisory panels and has been chairman of the Transportation Research Board's Committee on Hazardous Materials Transportation for more than 5 years. He holds an M.B.A. in transportation from the University of Oregon and a B.A. in economics from Western Maryland College.

Richard J. Ayen, a member of the NRC Committee on Review and Evaluation of Alternative Technologies for Demilitarization of Assembled Chemical Weapons (I and II), received his Ph.D. in chemical engineering from the University of Illinois. Dr. Ayen, now retired, was director of technology for Waste Management, Inc. He has extensive experience in the evaluation and development of new technologies for the treatment of hazardous, radioactive, industrial, and municipal waste. Dr. Ayen managed all aspects of Waste Management's Clemson Technical Center, including treatability studies and technology demonstrations for the treatment of hazardous and radioactive waste. His experience includes 20 years at Stauffer Chemical Company, where he was manager of the Process Development Department at Stauffer's Eastern Research Center. Dr. Ayen has published extensively in his fields of interest.

Robert A. Beaudet is chair of the NRC Committee on Review and Evaluation of Alternative Technologies for Demilitarization of Assembled Chemical Weapons (I and II). He received his Ph.D. in physical chemistry from Harvard University and has served on U.S. Department of Defense committees that address offensive and defensive chemical warfare. Dr. Beaudet was a member of the Army Science Board and chaired a committee that addressed chemical detection and trace gas analysis. He was chair of a series of Air Force technical workshops to develop master R&D plans for chemical warfare defense. He also served on the NRC's Committee on Chemical and Biological Sensor Technologies and its Committee on Energetic Materials and Science Technology. Most of his career has been devoted to research in molecular structure and molecular spectroscopy. Dr. Beaudet was a member of the Board on Army Science and Technology (BAST) and served as the BAST liaison to the Review and Evaluation of the Army Chemical Stockpile Disposal Program Committee during the development of nine reports. He is the author or coauthor of more than 100 technical reports and papers.

Lisa M. Bendixen is a principal in the environment and risk practice at Arthur D. Little, Inc. Since joining the company in 1980, Ms. Bendixen has been involved in risk management and risk assessment studies for numerous industries. She is the secretary of the NRC Transportation Research Board's Committee on Hazardous Materials and was the U.S. delegate to the International Electrotechnical Commission's working group on risk analysis until early 1999. She was a member of the NRC Committee on Fiber Drum Packaging for Transporting Hazardous Materials and is past chair of the Safety Engineering and Risk Analysis Division of the American Society of Mechanical Engineers. She has been involved in many studies on the chemical demilitarization of M-55 rockets, including the identification and quantification of failure modes leading to agent release based on a generic disposal facility design; evaluations of sources of risk in separating agent from energetic components in a rocket; and preparation of criteria for evaluating storage, transportation, and on-site disposal options. Ms. Bendixen earned an M.S. in operations research at the Massachusetts Institute of Technology.

Joan B. Berkowitz, managing director of Farkas Berkowitz and Company, has extensive experience in environmental and hazardous waste management and technologies for the cleanup of contaminated soils and groundwater and a strong background in physical chemistry and electrochemistry. She has contributed to several U.S. Environmental Protection Agency studies, has been a consultant on remediation techniques, and has assessed various destruction technologies. Dr. Berkowitz is the author of numerous publications on hazardous waste treatment and environmental subjects. She was a member of the NRC Panel on Review and Evaluation of Alternative Chemical Disposal Technologies and is currently a member of the NRC Committee on Review and Evaluation of Alternative Technologies for Demilitarization of Assembled Chemical Weapons (I and II). She has a Ph.D. from the University of Illinois in physical chemistry.

Judith A. Bradbury, technical manager at Battelle Patuxent River, is currently evaluating public involvement programs across the U.S. Department of Energy (DOE) complex. She previously completed a series of evaluations of the effectiveness of DOE's 12 site-specific advisory boards and led an assessment of community concerns about incineration and community perspectives on the U.S. Army Chemical Weapons Disposal Program. Dr. Bradbury is a member of the Risk Assessment and Policy Association. She earned a B.S. in sociology from the London School of Economics, an M.A. in public affairs from Indiana University of Pennsylvania, and a Ph.D. in public and international affairs from the University of Pittsburgh.

Martin C. Edelson has been a member of the staff at the Ames Laboratory since 1977 and is an adjunct associate professor of mechanical engineering at Iowa State University. His research interests include risk communication and the development of laser-based methods for materials processing and characterization. Dr. Edelson was a member of the Munitions Working Group and the DOE Laboratory Directors' Environmental and Occupational/Public Health Standards Steering Group. He currently represents the Ames Laboratory on the DOE Strategic Laboratory Council and the Subsurface Contamination Focus Area Lead Laboratory. He is also a liaison from the DOE Characterization, Monitoring, and Sensor Technology Crosscutting Program to the DOE Tanks Focus Area. Dr. Edelson was a technical editor of *Risk Excellence Notes*, a publication funded by the DOE Center for Risk Excellence, from 1998 to 2001, and is a member of the Executive Committee of ASME's Environmental Engineering Division. He earned a B.S. in chemistry and an M.A. in physical chemistry from the City College of New York and a Ph.D. in physical chemistry from the University of Oregon.

Sidney J. Green, a member of NAE, is chairman and chief executive officer of TerraTek, a geotechnical research and services firm in Salt Lake City focused on natural resource recovery, civil engineering, and defense problems. Before that, he worked at General Motors and the Westinghouse Research Laboratory. He has an extensive background in mechanical engineering, applied mechanics, materials science, and geoscience applications and is a former member of the NRC Geotechnical Research Board. He was named Outstanding Professional Engineer of Utah and is the recipient of the ASME Gold Medallion Award and the Lazan Award from the Society of Experimental Mechanics. Mr. Green received the degree of Engineer in engineering mechanics from Stanford University and an M.S. from the University of Pittsburgh and B.S. from the University of Missouri at Rolla, both in mechanical engineering.

Paul F. Kavanaugh, an engineering management consultant, was the director of government programs for Rust International, Inc., and director of strategic planning for Waste Management Environmental Services. During his military service, he served with the U.S. Army Corps of Engineers, DOE, and the Defense Nuclear Agency and managed engineering projects supporting chemical demilitarization at Johnston Atoll. He earned a B.S. in civil engineering from Norwich University and an M.S. in civil engineering from Oklahoma State University. Brigadier General Kavanaugh is a fellow of the Society of American Military Engineers.

Todd A. Kimmell is principal investigator in the Environmental Assessment Division at Argonne National Laboratory. He is an environmental scientist and policy analyst. Mr. Kimmell was selected for membership on the committee for his expertise as an environmental regulatory and permitting specialist with more than 20 years of extensive experience in solid and hazardous waste management, program and policy development, chemical munitions and explosives waste, as well as in many other activities related to regulatory and permitting issues. He graduated from the George Washington University with a master's degree in environmental science.

Douglas M. Medville retired from MITRE as program leader for chemical materiel disposal and remediation. He has led many analyses of risk, process engineering, transportation, and alternative disposal technologies and has briefed the public and senior military officials on the results. Mr. Medville led the evaluation of the operational performance of the Army's chemical weapon disposal facility on Johnson Atoll and directed an assessment of the risks, public perceptions, environmental aspects, and logistics of transporting recovered non-stockpile chemical warfare materiel to candidate storage and disposal destinations. Before that, he worked at Franklin Institute Research Laboratories and General Electric. Mr. Medville earned a B.S. in industrial engineering and an M.S. in operations research, both from New York University.

Winifred G. Palmer is a consultant in toxicology. She was a toxicologist for the U.S. Army between 1989 and 2000 for the Biological Research and Development Laboratory and the Center for Health Promotion and Preventive Medicine. Her work for the Army included assessment of health risks associated with chemical warfare agents, development of a military field water quality standard for the nerve agent BZ, development of the military standard for fog oil (an obscurant smoke), and studies of the mutagenicity of rifle emissions and the bioavailability of TNT in composts of TNT-contaminated soils. Dr. Palmer is a member of the Society of Toxicology, and her numerous publications span more than two decades of work in the field. She has a Ph.D. in biochemistry from the University of Connecticut and a B.S. in chemistry and biology from Brooklyn College.

George W. Parshall, a member of NAS, graduated from the University of Illinois with a Ph.D. in organic chemistry. After nearly 40 years of service, Dr. Parshall retired from E.I. duPont de Nemours and Company as director of chemical science in the Central Research and Development Division. He was selected for membership on this committee because of his experience in organic and inorganic chemistry and catalysis and in conducting and supervising chemical research. Dr. Parshall is a past member of the NRC Board on Chemical Science and Technology and the NRC Committee on Review and Evaluation of the Army Chemical Stockpile Disposal Program.

James P. Pastorick is president of Geophex UXO, Ltd., an unexploded ordnance (UXO) consulting firm based in Alexandria, Virginia, that specializes in UXO planning and management consulting to state and foreign governments. Since he retired from the U.S. Navy as an explosives ordnance disposal officer and diver in 1989, he has been working on civilian UXO clearance projects. Prior to starting his present company, he was the senior project manager for UXO projects at UXB International, Inc., and the IT Group.

R. Peter Stickles graduated from Northwestern University with a Master of Science in chemical engineering. For 8 years he was a process engineer with Stone & Webster Engineering Co. He retired after 27 years of service from Arthur D. Little, Inc., as a principal in Global Environmental and Risk Practice. Mr. Stickles was selected for membership on this committee for his more than 35 years of experience in a variety of activities in the area of chemical process engineering, including development and project activities that specified the design of petrochemical plants based on the thermal cracking of hydrocarbons and participation in the design and startup of plants to produce ethylene and alpha-olefins.

William J. Walsh is an attorney and partner in the Washington, D.C., office of Pepper Hamilton LLP. Prior to joining Pepper, he was a section chief in the EPA Office of Enforcement. His legal experience encompasses environmental advice and environmental injury litigation involving a broad spectrum of issues pursuant to a variety of environmental statutes, including the Resources Conservation and Recovery Act (RCRA) and the Toxic Substances Control Act (TSCA). He represents trade associations, including the Biotechnology Industry Organization, in rulemaking and other public policy advocacy; represents individual companies in environmental actions (particularly in negotiating cost-effective remedies in pollution cases involving water, air, and hazardous waste); and advises technology developers and users on taking advantage of the incentives for, and eliminating the regulatory barriers to, the use of innovative environmental technologies. He previously served on NRC committees concerned with Superfund and RCRA corrective action programs and the use of appropriate scientific groundwater models in environmental regulatory programs and related activities. Mr. Walsh holds a J.D. from George Washington University Law School and a B.S. in physics from Manhattan College.

Ronald L. Woodfin is a recently retired staff member of Sandia National Laboratories, where he coordinated work on mine countermeasures and demining, including sensor development. He is currently an adjunct professor of mathematics at Wayland Baptist University, Albuquerque Campus. Previously, he worked at the Naval Weapons Center, Naval Undersea Center, and Boeing Commercial Airplane Division. Dr. Woodfin has been an invited participant at several international conferences on demining and has served on an advisory task force on humanitarian demining for the General Board of Global Ministries of the United Methodist Church. He also serves as pastor of Cedar Crest Baptist Church, Cedar Crest, New Mexico. Dr. Woodfin earned a B.S. in aerospace engineering from the University of Texas and an M.S. in aeronautics and astronautics and a Ph.D. in engineering mechanics from the University of Washington. He recently completed service on the Committee for Mine Warfare Assessment of the Naval Studies Board of the National Research Council.

B

Committee Meetings and Other Activities

MEETINGS

First Committee Meeting, January 22-24, 2001, National Research Council, Washington, D.C.

Presentations:

Opening Remarks
James Bacon, Program Manager, Chemical Demilitarization
Margo Robinson, Budget Manager, ASAALT

Non-Stockpile Chemical Materiel Program Update and Status
William Brankowitz, Deputy to the Product Manager, Non-Stockpile Chemical Materiel Product

Second Committee Meeting, March 15-16, 2001, Edgewood, Maryland, and Aberdeen Proving Ground, Maryland.

Presentations:

Product Manager's Status Briefing
Christopher M. Ross, Product Manager, Non-Stockpile Chemical Materiel Product

Committee Subgroup Meetings
John Nunn, Maryland Citizens' Advisory Commission
Ken Stachew, Remediation Advisory Board

U.S. House of Representatives Armed Services Committee Concerns
Jean Reed, House Armed Services Committee

Third Committee Meeting, May 23-24, 2001, Washington, D.C.

Presentation:

U.S. Army Non-Stockpile Chemical Materiel Product (NSCMP) Overview/Status
William Brankowitz, Deputy to the Product Manager, Non-Stockpile Chemical Materiel Product

Fourth Committee Meeting, July 10-11, 2001, Aberdeen, Maryland

Presentations:

Opening Remarks
James Bacon, Program Manager for Chemical Demilitarization

Remarks
Henry Dubin, Acting Deputy Assistant Secretary of the Army for Chemical Demilitarization

Non-Stockpile Chemical Materiel Product Update
Christopher M. Ross, Product Manager, Non-Stockpile Chemical Materiel Product

Committee Subgroup Meetings
Don Benton, Munitions Assessment and Processing System
Jeff Harris, Rapid Response System
Dave Hoffman, Explosive Destruction System
Eric Kauffman, Pine Bluff Non-Stockpile Facility

79

Committee writing meeting, August 28, 2001, Washington, D.C.

Fifth committee meeting, September 25-26, 2001, Edgewood, Maryland

Presentations:

U.S. Army Non-Stockpile Chemical Materiel Product (NSCMP) Update
Christopher M. Ross, Product Manager, Non-Stockpile Chemical Materiel Product

Destruction of Chemical Warfare Materials in Albania Using Plasmox© Technology
Joseph Sudol, Burns and Roe

Technology Test Program for Treatment of NSCMP Feeds
Joseph Cardito, Stone and Webster, and
Edward Doyle, Program Manager for Chemical Demilitarization

Sixth committee meeting, November 8-9, 2001, Irvine, California

Presentations:

View from the Pentagon
Henry Dubin, Acting Deputy Assistant Secretary of the Army for Chemical Demilitarization and Assistant Secretary of the Army for ALT

Product Manager's Product Status Briefing
Christopher M. Ross, Product Manager, Non-Stockpile Chemical Materiel Product

Committee writing meeting, November 28, 2001, Washington, D.C.

Seventh committee meeting, March 27-28, 2002, Washington, D.C.

Presentation:

Product Manager's Product Status Briefing
Christopher M. Ross, Product Manager, Non-Stockpile Chemical Materiel Product

SITE VISITS

Core Group Meeting, Aberdeen Proving Ground, Maryland, June 12, 2001

Site Team
Judith A. Bradbury, committee member
Martin C. Edelson, committee member

Donovan Blast Chamber Demonstration, Brussels, Belgium, July 25, 2001

Site Team
George W. Parshall, committee member

Restoration Advisory Board Meeting, Aberdeen Proving Ground, Maryland, July 28, 2001

Site Team
Judith A. Bradbury, committee member

Pine Bluff Arsenal, Arkansas, August 9, 2001

Site Team
Judith A. Bradbury, committee member
Martin C. Edelson, committee member
Nancy T. Schulte, study director

Pine Bluff Non-Stockpile Facility, October 13, 2001

Site Team
Judith A. Bradbury, committee member
Martin C. Edelson, committee member

PMCD Environmental Forum, Atlanta, Georgia, October 18, 2001

Site Team
Todd Kimmell, committee member

C

Evaluation of the Suitability of Stockpile Chemical Disposal Facilities for Treating Stored Non-Stockpile CWM

STOCKPILE DISPOSAL FACILITIES

Aberdeen Proving Ground

The chemical disposal facility (CDF) at Aberdeen Proving Ground (APG), Maryland, is designed to process only the stockpile materiel located at APG. These are ton containers filled with mustard (H/HD). No explosives are involved. As of November 2001, the Army's Project Manager for Alternative Technologies and Approaches proposed an expedited disposal of the stockpile of these mustard-filled ton containers. Under the proposed approach, the agent would be drained from ton containers using a vacuum system rather than punching and draining, the agent removal would be done in a glove box in an existing building, agent would be neutralized, the ton containers rinsed, and the neutralent sent to a commercial posttreatment facility rather than be treated on-site by biotreatment. The Army expects that the expedited process would allow all the agent in the stockpile ton containers to be destroyed by the fall of 2002 (U.S. Army, 2001).

In addition to the stockpile ton containers, there are 125 non-stockpile items at APG, of which 28 contain mustard (HD, HT, HS). The following are the non-stockpile mustard items:

Munitions (9 items)
  6 75-mm projectiles (contain explosive)
  1 4.2-inch mortar round (contains explosive)
  2 4-inch mortar rounds (do not contain any explosive; to be used in APG testing)

Chemical Sample Containers (19 items)
  5 30-gallon drums containing metal "pumpkins" of agent
  3 5-pint cans containing bottles or vials of agent
 11 55-gallon drums containing metal "pumpkins" of agent

Although the Aberdeen Chemical Agent Disposal Facility (ABCDF) is equipped to monitor for HD, the facility's hardware configuration does not permit it to process some of the items listed above. The exceptions are the larger containers such as the 30-gallon and 55-gallon chemical sample drums, which may be opened using the same punch-and-drain station used to access agent in the ton containers. The ABCDF is not equipped to process the nine projectiles or mortar rounds since it will not have the demilitarization machines needed to remove energetics (the projectile mortar disassembly machine (PMD)) or to drain agent (the multipurpose demilitarization machine (MDM)) in an explosive containment room.

In addition to the items containing mustard, the following 97 non-stockpile items are at APG:

Chemical Sample Empty Containers (96) consisting of:
  5 GB 30-gallon drums
  6 GB multipack bottles, vials
  2 GB ton containers (these have been disposed of[1])
 12 GB steel cylinders
 10 L   multipack bottles and vials

---

[1]Christopher Ross, PMNSCM, briefing to the committee on July 9, 2001.

55 VX drums, cans, buckets
 6 VX DOT bottles

Munition (1 item)
 1 CG 4.2-inch mortar (explosive)

The chemical sample containers can be disposed in the Aberdeen Chemical Transfer Facility (CTF), an R&D facility at APG that has processed munitions, sample bottles, and ton containers containing a variety of chemical fills. The CTF does not have a capability for processing explosively configured munitions but does contain a chemical agent transfer system that can drain ton containers. There is no treaty-imposed time limit on operation of the CTF, and if its schedule permits, it can dispose of the container items listed above. The Product Manager for Non-Stockpile Chemical Materiel (PMNSCM) has proposed using the CTF to destroy appropriate NSCWM items found at APG.

The CTF will perform the neutralization of agent accessed using the MAPS at APG (see Appendix D). This alternative is favored by the local Citizens' Advisory Commission.

Anniston Chemical Activity

The stockpile materiel stored at Anniston Chemical Activity consists of GB projectiles and rockets, HD projectiles and ton containers, and VX projectiles, mines, and rockets.

Also stored at Anniston are 133 non-stockpile chemical sample containers, consisting of the following:

  2 GB ton containers
119 GB vials
  5 HD DOT bottles
  7 VX DOT bottles

If a permit modification can be obtained, the two GB-filled ton containers can be handled in the Anniston Chemical Disposal Facility (ANCDF) since that facility can monitor for GB and has the equipment to punch and drain ton containers. The bottles and vials are more problematic. If they can be opened and placed in a tray, they can be fed into the metal parts furnace (MPF) or possibly into the deactivation furnace system (DFS). If it is necessary to access the agent in the bottles, they can be crushed prior to feeding into the DFS. The PMNSCM proposes to use the ANCDF for disposal of the Anniston Army Depot chemical samples. It is also possible to destroy agents in the vials and DOT bottles with the EDS-1, although this may not be as economical as using the ANCDF.

Bluegrass Army Depot

Chemical stockpile items stored at Bluegrass consist of mustard-filled projectiles, GB projectiles and rockets, and VX projectiles and rockets. There are only four non-stockpile chemical sample containers stored at Bluegrass:

 2 HD DOT bottles
 1 GB ton container
 1 VX DOT bottle

Regardless of the technology selected, the Bluegrass CDF will not be equipped to handle the GB ton container since there are no ton containers in the chemical stockpile at Bluegrass Army Depot. The ton container contents can be transferred to DOT bottles for destruction in an EDS-1 if one is brought to Bluegrass Army Depot. Although the PMNSCM proposes to dispose of the chemical samples in the Bluegrass CDF, a chemical disposal technology has not been selected for this location, nor is the ability of such a facility to dispose of the four items listed above known.

Deseret Chemical Depot

The chemical stockpile at Deseret consists of mustard-filled projectiles and ton containers; GB-filled projectiles, rockets, bombs, and ton containers; VX-filled projectiles, mines, rockets, spray tanks, and ton containers; and lewisite-filled ton containers.

For purposes of discussion and categorization, the known inventory of non-stockpile materiel at both Deseret Chemical Depot and the nearby Dugway Proving Ground is grouped into three categories. These are listed in order of decreasing compatibility with the Tooele Chemical Disposal Facility (TOCDF) at Deseret Chemical Depot:

Same Fill as Stockpile Inventory, Same Items (18 items)
 1 HD   ton container (has been disposed of)
10 HD   4.2-inch mortar (explosive)
 2 HD   105-mm (explosive)
 1 GB   155-mm (explosive)
 2 GB   155-mm (nonexplosive)—one is empty but contaminated
 2 HD   4.2-inch mortar (nonexplosive)

Same Fill as Stockpile Inventory, Different Items (139 items)
45 HD, HT   chemical samples (miscellaneous containers)
 1 GB   ampoule
 1 GB   155-mm T77 (explosive)
 1 GB   6-inch round (explosive)
 1 GB   M-125 bomblet (explosive)
 1 GB   M-139 half bomblet (nonexplosive)
48 GB, GD   chemical sample bottles
 8 HD, HT   chemical sample bottles
28 VX   chemical sample bottles
 5 VX (EA-1699) chemical sample bottles

Different Fill (Lewisite) from Stockpile Inventory (17 items)
12 L   4.2-inch mortar rounds (containing explosive)
 4 L   105-mm projectiles (containing explosive)
 1 L   chemical sample bottle

Of the 174 non-stockpile items listed above, 19 contain the same fills as the items processed in the TOCDF (HD, GB, VX) and either are identical to the stockpile items or have configurations that enable them to be processed in the stockpile demilitarization machines after modifications are made. The remaining 155 items have a lewisite fill (17 items), or are nonstandard in size (3 items), or are small chemical sample containers (135 items). The PMNSCM proposes to dispose of the items having a lewisite fill at the Army's Chemical Agent Munitions Disposal System (CAMDS) facility at Deseret Chemical Depot since that facility is able to process and monitor for this agent. The non-stockpile items containing mustard and the GB and VX nerve agents are to be destroyed in the TOCDF if permitting and public approval are obtained. There is some opposition to any further use of the TOCDF following the completion of stockpile destruction. The TOCDF is scheduled to complete operations during the fourth quarter of FY 2003; thus, there is sufficient time available for processing non-stockpile items after the stockpile campaigns are completed rather than interspersing them with similar stockpile items having the same fills.

Of the 173 non-stockpile items listed above (excluding the mustard-filled ton container), 37 are candidates for processing in an EDS:

10 HD 4.2-inch mortars (explosive)
 2 HD 105-mm (explosive)
 1 GB 155-mm (explosive)
 2 GB 155-mm (nonexplosive)
 2 HD 4.2-inch mortars (nonexplosive)
 1 GB T77 155-mm (explosive)
 1 GB 6-inch projectile (explosive)
 1 GB M125 bomblet (explosive)
 1 GB M139 half bomblet (nonexplosive)
12 L   4.2-inch mortars (explosive)
 4 L   105-mm projectiles (explosive)

Of these 37 items, the first 18 can be processed in the TOCDF since they are similar to the stockpile munitions being disassembled and destroyed in that facility. The GB-filled 6-inch projectile, bomblet, and half bomblet differ from the stockpile munitions and would require modifications to TOCDF's demilitarization machines prior to processing. The 16 munitions having a lewisite fill cannot be processed in the TOCDF since that facility is not permitted to destroy lewisite. The remaining non-stockpile items at DCD are all chemical sample containers containing agent and can be destroyed in the RRS (size permitting) or in an EDS.

All of these items can also be processed at the Army's CAMDS facility, at Deseret. CAMDS operates under a RCRA operating permit and, as a non-stockpile disposal facility, is not affected by the 2007 CWC treaty deadline. CAMDS also has other missions and tests to conduct, however, and completing the destruction of the non-stockpile items listed above before 2007 will require some scheduling prioritization.

Chemical Agent Munitions Disposal System

CAMDS, located at Deseret Chemical Depot near Tooele, Utah, is the Army's R&D facility for building and testing prototype chemical demilitarization hardware and processes. The demilitarization machines used in the stockpile chemical disposal facilities and prototypes of the incinerators, for example, were fabricated and tested at CAMDS. CAMDS has been used by the non-stockpile product manager to develop, assemble, and test the RRS used for the disposal of CAIS. CAMDS has also been used to test systems for the biological degradation of chemical agents and is currently the Army's facility for the disposal of chemical materiel containing the arsenical agent lewisite. The lewisite in stockpile ton containers and non-stockpile items containing lewisite (mortar rounds, projectiles, and a chemical sample bottle) are intended for destruction at CAMDS.

The CAMDS physical facility consists of several buildings, incinerators, and engineering offices. It is a valuable facility that can undertake specialized projects, destroy relatively small quantities of chemical agents, and develop and test equipment used for chemical munitions disposal.

Newport Chemical Depot

There is no non-stockpile materiel stored at Newport. The Army may find some VX in pipes during demolition of the old production facility, but this would be treated as stockpile materiel. Thus, there is no need to consider the Newport CDF for NSCWM.

Pine Bluff Arsenal

The chemical stockpile at Pine Bluff consists of mustard-filled (HD and HT) ton containers, GB-filled rockets, and VX-filled rockets and mines.

There are also 69,878 non-stockpile items stored at Pine Bluff, by far the largest part of the non-stockpile inventory. For discussion and categorization purposes, the known inventory of non-stockpile materiel found at Pine Bluff is grouped below into six categories and listed in order of decreasing compatibility with the Pine Bluff CDF.

Same Fill as Stockpile, Same or Similar Items (11 items)
  2 GB   ton containers
  9 HD   M70A1 bombs

The GB-filled ton containers can be co-processed with stockpile GB rockets in the PBCDF, and, indeed, the PMNSCM proposes to do this. The bombs, if not explosively configured, can be drained using the same punch-and-drain equipment used for the ton containers. Agent would be destroyed in the liquid incinerator and the bomb bodies brought to the 5X level in the metal parts furnace.[2] If the bombs do contain explosives, use of the stockpile CDF may be inappropriate if a larger explosion containment room than required for stockpile operations is needed.

Same Fill as Stockpile, Different Items (763 items)
  727 HD   4.2-inch mortars (explosive)
   16 HD   75-mm projectiles (explosive)
   12 HD   200-mm Livens projectiles (explosive)
    1 HD   155-mm projectile (explosive)
    3 HD   75-mm projectiles (nonexplosive)
    1 HD   105-mm projectile (explosive)
    1 HD   4-inch cylinder (nonexplosive)
    2 VX   chemical sample containers

The explosively configured items are not compatible with the PBCDF since the facility will not have the demilitarization machines (the projectile mortar disassembly machine and the multipurpose demilitarization machine) that are needed to remove energetics from the projectiles and mortar rounds and to drain agent from them. The nonexplosively configured items could be processed in the Pine Bluff CDF, although there would be a need to access the agent, perhaps by modifying the bulk container handling system (punch and drain) or by drilling into the items and draining the agent. The PMNSCM proposes to destroy the H and VX chemical sample containers in the PBCDF, subject to permits and public acceptability.

Different Fill, Different Items (483 items), Poor Compatibility with Pine Bluff CDF
    1 CG   4.2-inch mortar (explosive)
    3 CG   200-mm Livens projectiles (explosive)
  479 HN   150-mm Traktor rockets with warheads[3]

CG (phosgene) is a gaseous fill in the projectiles and mortar rounds. Major changes in processing would be needed to collect it, contain it, transfer it to the incinerator, and incinerate it. In addition, equipment would have to be modified to monitor for CG.

The World War II German Traktor rockets are also not suitable for processing in the Pine Bluff CDF. They are more complex and heavier than the M-55 rockets. The plant's monitors are not set up to monitor for nitrogen mustard (HN) and would have to be recalibrated. Destroying the HN in the liquid incinerator would also require a permit modification and trial burn. The time needed to do these things, especially in light of the limited time available (6 months) for non-stockpile operations following completion of the stockpile campaigns near the end of 2006, is too long to allow these items to be destroyed in the Pine Bluff CDF. The Army plans to destroy the Traktor rockets in the explosion containment room in the Pine Bluff Non-Stockpile Facility (PBNSF). This facility will be constructed and used at Pine Bluff Arsenal for disposing of much of the non-stockpile inventory stored there.

Items with Lewisite Fill, Excluding CAIS (4,378 items)
      2   4.2-inch mortars (explosive)
      1   chemical sample vial
  4,375   ton containers (empty but once contained L)

In sampling the ton containers in 1995, stable forms of lewisite, arsenic, and mercury were detected, indicating that the ton containers had contained lewisite in the past. Although from a mechanical standpoint the empty ton containers can be processed in the plant's bulk container handling facility, the plant is not set up to process items containing arsenic or mercury. The empty ton containers at Pine Bluff will be decontaminated to a $3X^4$ condition in a decontamination enclosure and will then be sent to Rock Island Arsenal for smelting.

CAIS Items, All Fills (7,120 items)

CAIS items at Pine Bluff Arsenal contain many fills (HD, L, PS, CG, CN, DM, CK, and HN). The 5,814 CAIS con-

---

[2]Treatment of solids to a 5X decontamination level is accomplished by holding a material at 1,000 °F for 15 minutes. This treatment results in completely decontaminated material that can be released for general use or sold (e.g., as scrap metal) to the general public in accordance with applicable federal, state, and local regulations.

[3]In presentations to the committee, the PMNSCM indicated that some of these Traktor rockets contained arsenical agents.

[4]"3X" refers to the level at which solids are decontaminated to the point that agent concentration in the headspace above the encapsulated solid does not exceed the health-based, 8-hour, time-weighted average limit for worker exposure. The level for mustard agent is 3.0 micrograms/m^3 in air. Materials classified as 3X may be handled by qualified plant workers using appropriate procedures but are not releasable to the environment or for general public reuse. In specific cases in which approval has been granted, a 3X material may be shipped to an approved hazardous waste treatment facility for disposal in a landfill or for further treatment.

taining various forms of mustard can be processed by feeding the CAIS directly into either the deactivation furnace system or the metal parts furnace (MPF), or the vials and ampoules can be crushed first. The industrial chemicals in the CAIS could also be processed in the deactivation furnace system or the MPF, although the need to monitor for these fills, recalibrate the monitors, obtain permit modifications, and conduct possible trial burns makes this application less desirable than processing CAIS containing stockpile fills. CAIS containing lewisite (L) and adamsite (DM) would not be processed in the Pine Bluff CDF since the plant is not equipped to handle or monitor for fills containing arsenic. As an alternative to using the Pine Bluff CDF, all of the agent-containing CAIS could be destroyed in an RRS if one is brought to or located at PBA. CAIS vials containing industrial chemicals may be processed using commercial incinerators, as is planned by the NSCMP.

Binary Chemical Warfare Materiel Components (57,123 items)

56,820	DF	M-20 containers
7	DF	55-gallon drums
293	QL	55-gallon drums
3	QL	containers (boxes, cans)

Use of the liquid incinerator at the Pine Bluff CDF to destroy the DF will be impractical since there is insufficient time to process the large number of DF containers. Also, depending on operating conditions, the fluorine in the DF could erode the liquid incinerator's refractory brick and mortar and require frequent rebricking of the liquid incinerator. It may be possible to process the QL drums and containers in the liquid incinerator, but they can be processed more easily in the same separate facility that will process DF. As part of its technology test program, the NSCMP is considering several options for the disposal of binary chemical warfare materiel components, including the use of plasma arc technology and chemical neutralization.

Pueblo Chemical Depot

The chemical stockpile at Pueblo consists entirely of mustard-filled projectiles. There are 12 non-stockpile mustard-filled chemical sample containers (DOT bottles) stored at Pueblo.

Selection of a technology at Pueblo Chemical Depot for the Pueblo CDF has not been made. Candidates include (1) a modified baseline process in which agent-filled projectiles are treated in a four-zone MPF and uncontaminated energetics are disposed at an off-site facility; (2) and (3) two ACWA alternative chemical technologies where the munitions are disassembled by a modified baseline process, the mustard is treated with water to hydrolyze it, and the energetics are treated with caustic to hydrolyze them. The products of the hydrolysis are then exposed to biotreatment or SCWO. (One of the two ACWA alternative technologies also uses cryofracture to expose the agent in the projectiles). All of these processes can handle the DOT bottles, although the demilitarization machines will need some modification in order to access agent in the baseline and modified baseline processes. The steel DOT bottles could also be crushed in the cryopress after cooling in the cryobath.

With only 12 DOT bottles and more than 780,000 stockpile projectiles at Pueblo, destroying the bottles should not have any impact on plant operations, so the PMNSCM proposes to destroy these non-stockpile items in the Pueblo CDF.

Umatilla Chemical Depot

The chemical stockpile at Umatilla consists of HD ton containers, GB bombs, projectiles, and rockets, and VX projectiles, rockets, mines, and spray tanks. There are only five non-stockpile items stored at Umatilla:

4	GB	ton containers
1	VX	ton container

Since the Umatilla CDF is designed to process the stockpile ton containers and can monitor for GB and VX, it can process the five non-stockpile ton containers as well. This can be done as part of a coprocessing operation when the stockpile GB and VX rockets are being destroyed since agent monitors will be detecting GB and VX. The existing permit allows all CWM at Umatilla to be destroyed, including NSCWM, but a permit modification will be needed to include five non-stockpile ton containers. The state of Oregon prefers that the Army destroy the non-stockpile ton containers in the CDF to save the time and effort of writing a new permit for a transportable unit, and the PMNSCM proposes to take this action.

REFERENCE

U.S. Army. 2001. U.S. Army Chemical Demilitarization Program Releases. Updated Official Schedule and Cost Estimates, Press Release. Information Office, October 4. Aberdeen Proving Ground, Md.: Program Manager for Chemical Demilitarization.

D

Non-Stockpile Facilities

This appendix reviews the two non-stockpile facilities proposed for treatment of NSCWM, the Munitions Assessment and Processing System (MAPS), under construction at Aberdeen Proving Ground, Maryland, and the Pine Bluff Non-Stockpile Facility (PBNSF), proposed for Pine Bluff Arsenal, Arkansas.

MUNITIONS ASSESSMENT AND PROCESSING SYSTEM

The Munitions Assessment and Processing System (MAPS) is a facility developed specifically to destroy explosively configured non-stockpile chemical munitions at the Aberdeen Proving Ground (APG). MAPS has five components: a negative pressure filtration system, an air monitoring system, a glove box, an explosive containment chamber, and a burster detonation vessel. A plan of the facility showing the location of the key components is seen in Figure 2-1.

Description of Main Components

Negative Pressure Filtration System

The filtration system contains carbon filters for removing the contaminants from the building air. This system also maintains the process areas in the facility under negative pressure. Negative pressure is an engineering control that ensures that potentially contaminated building air is not released into the environment.

Air Monitoring System

MAPS contains an array of 15 minicams that monitor the facility air for the presence of chemical agents. These monitors are designed to alert the operators to chemical agent leaks so that appropriate action can be implemented.

Glove Box

The glove box is located in the process room and provides a means of handling the recovered munitions under negative pressure to reduce the risk of chemical agents being released into the process room.

Explosive Containment Chamber

This item is also located in the process room. The explosive containment chamber (ECC) is where the munitions will be either cut or drilled. In the event of an accidental detonation during the cutting or drilling, the ECC is designed to withstand a blast of up to 13 pounds of TNT without a vapor release but is expected to be used to destroy projectiles not exceeding 5 pounds of TNT equivalent.

Burster Detonation Vessel

The burster detonation vessel (BDV) is a commercial detonation vessel used to destroy the empty munition body, fuze, and burster. The BDV contains both the blast overpressure and the metal fragments. The overpressure will be vented to the negative pressure filtration system.

Description of Process

Before any chemical munitions are processed in MAPS, they must be properly identified and their contents verified using approved radiography and nonintrusive characterization technologies. Only after this process has been completed and the chemical munitions are considered safe will they be processed in the MAPS. MAPS is designed to destroy projectiles as large as 155 mm or having a maximum amount of explosive not to exceed 5 pounds of TNT equivalent, which provides a margin of safety.

When munitions that have been declared safe arrive at MAPS to be processed, they move from the processing room and are placed in the glove box. Any packing material that is around the munitions is removed, and the munitions are then placed onto a drill or cut box. The drill or cut box is then moved to the explosive containment vessel, where the munitions' chemical fill will be removed by either cutting or drilling a hole into the munitions large enough to allow the chemical fill to drain out. The chemical fill is drained into an approved shipping container that is sent to the chemical transfer facility at APG to be neutralized. The munitions' bodies are decontaminated and then transported to the MAPS burster detonation unit, where they are detonated. After checking to ensure that the metal debris meets the 3X standard of decontamination, the metal is sent to an appropriate permitted landfill.

PINE BLUFF NON-STOCKPILE FACILITY

The Pine Bluff Non-Stockpile Facility (PBNSF) is currently in the design phase, which is scheduled to run during fiscal years 2001 and 2002. This facility will be designed as a site-specific solution to process the recovered chemical warfare materiel (RCWM), binary chemical weapon components, chemical agent identification sets (CAIS), and chemical samples at Pine Bluff Arsenal. It will have permanent walls, ceiling, and roof, but due to public concerns that this facility might be used to process non-stockpile items not currently stored at Pine Bluff, the equipment inside the building will be modular, skid-mounted, and easily transportable to permit removal after operations have been completed. The non-stockpile inventory at Pine Bluff is listed in Table 1-1. If the workload is broken down by item category, binaries account for 81.7 percent; CAIS, for 10.2 percent; empty ton containers, for 6.3 percent; munitions, for 1.8 percent; and chemical sample containers, for less than 1 percent. The construction of the facility is scheduled to begin in January 2004, with completion by July 2005.

Design Concept

The facility will be designed to operate 8 hours a day, 5 days a week. At this time no specific rate has been determined for the disposal of the binaries, the CAIS items, or the chemical samples. The facility will include a munition unpacking chamber, two ECCs, a neutralization trailer, soak tanks, and a secondary waste processor. The secondary waste processors under primary consideration are supercritical water oxidation and plasma arc, with the Army favoring plasma arc at this writing. Outside the walls of the PBSNF will be a detonation chamber, an EDS, and an RRS. When completed and operational, the facility will be able to destroy all items currently stored at Pine Bluff Arsenal.

Process Flow for RCWM

Before any of the munitions are delivered to the facility, they will have been previously assessed by the portable isotopic neutron spectroscopy (PINS) technique, x-rayed where appropriate, and the results reviewed by the Munitions Assessment and Review Board (MARB) to properly identify the contents. Data indicating the presence or absence of explosive and fuze components and the type of chemicals will be provided to the PBNSF operator for each item. The munitions will arrive at the facility overpacked individually in airtight containers. The following briefly describes the typical process flow for the categories of RCWM to be processed.

Munitions Having Unstable Energetics

The overpacked munitions are removed from storage and transported to the Munitions Warming Area, if necessary, or to the Unpack Area. After visual inspection of the overpack, the munitions are removed from the overpack and a visual inspection is performed. The munitions are then placed in a cradle and an x ray is taken to verify the contents. If a munition has an armed fuze or is determined to have unstable energetics, it is sent outside the facility to the detonation chamber (in this case, an Explosive Destruction System (EDS)) to be processed. In the EDS, a linear shaped charge is placed on the munitions to cut them open and expose their contents for chemical treatment. Conical shaped charges are used to detonate the internal energetic compounds. Once sealed inside the blast chamber, the charges are detonated, destroying most of the energetics and the CW material. The blast chamber is then loaded with neutralizing reagent and the contents are chemically neutralized. The neutralent is sent for secondary waste processing through PBNSF. The chamber is then flushed out with approved rinsates, with the liquid going to secondary treatment as necessary. The metal is immersed in a soak tank, where it is treated to a $3X^1$ condition before final disposal.

---

[1]"3X" refers to the level at which solids are decontaminated to the point that agent concentration in the headspace above the encapsulated solid does not exceed the health-based, 8-hour, time-weighted average limit for worker exposure. The level for mustard agent is 3.0 micrograms/m^3 in air. Materials classified as 3X may be handled by qualified plant workers using appropriate procedures but are not releasable to the environment or for general public reuse. In specific cases in which approval has been granted, a 3X material may be shipped to an approved hazardous waste treatment facility for disposal in a landfill or for further treatment.

Munitions Having Stable Energetics

Munitions having stable energetics are treated inside the facility. The munitions are transported to the ECC, where a remotely controlled device punches a hole into the munition, allowing the chemical to drain out. After draining, the ECC is rinsed and the liquid stream produced is sent to the chemical processing trailer, where the primary neutralization operations take place. The liquid waste is then sent to secondary treatment within PBNSF for final treatment. After the munition casing has been adequately rinsed, it is moved to the soak tank for further decontamination. When the munition has been verified to be in a 3X condition, it is loaded into the detonation chamber, located outside the facility, where destruction of the explosive components is accomplished. The detonation chamber can also be used for small NSCM items such as miscellaneous fuzes. The metal is then processed for final disposal, as shown in Figure D-1.

Nonexplosively Configured Munitions

The process for handling nonexplosively configured munitions is very similar to that for handling explosively configured munitions. The current design instructions call for the munitions to be processed in either the ECC or in the Chemical Agent Transfer System (CHATS), which will be designed for operations in a nonenergetic atmosphere. In either case the munition is breached using a remotely controlled operation, the chemical fill verified, and the chemicals drained from the munition are transferred to a secondary treatment facility for final treatment. The main difference is that after the munition is drained, properly rinsed, and soaked, it is taken to a cutting station, where it is cut into various sections. The sections are inspected to ensure they are in 3X condition before being released for disposal. The typical process flow for nonexplosively configured munitions is shown at Figure D-2.

Process Flow for Binaries

The final decision on how to process the binary chemicals has not been made. It is anticipated that binaries will be processed either by a punch-and-drain operation followed by neutralization or by feeding the DF canisters into a system—plasma arc, for example—in a single step.

Process Flow for CAIS

The Rapid Response System (RRS) will be set up as part of the PBNSF operation but will be located outside the PBNSF structure and will be used to process all CAIS items. The overpack containing the CAIS is removed from storage and transported directly to the RRS. The RRS contains a series of linked glove boxes designed to remove the CAIS ampoules and bottles from their packages, identify their con-

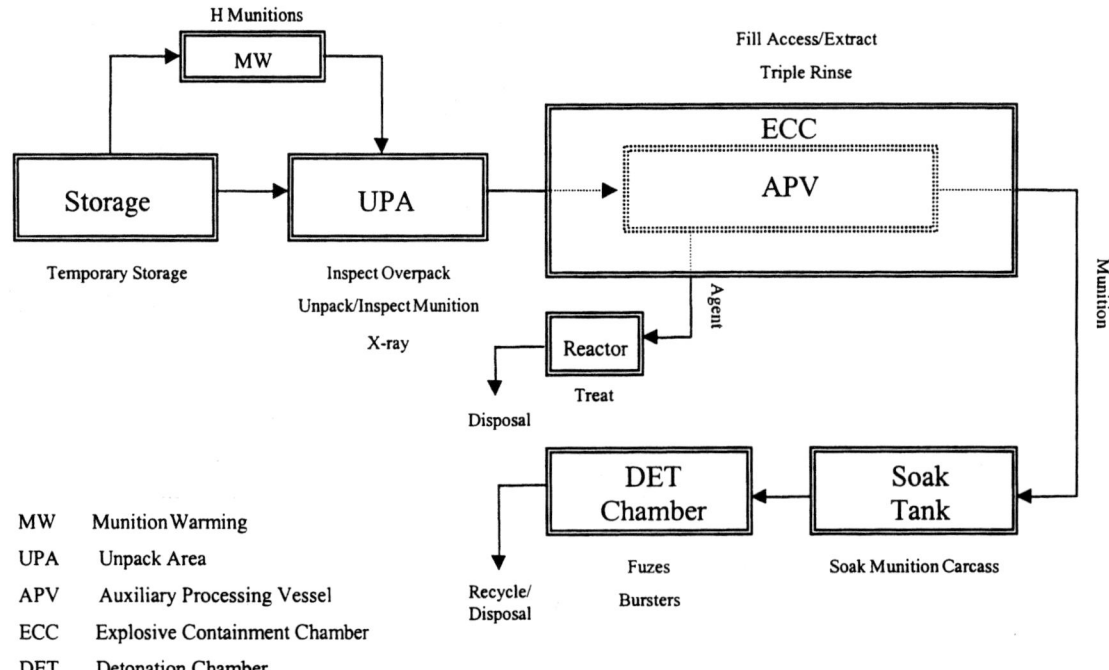

FIGURE D-1 Typical process flow for explosively configured munitions at PBNSF. SOURCE: U.S. Army (2000).

APPENDIX D

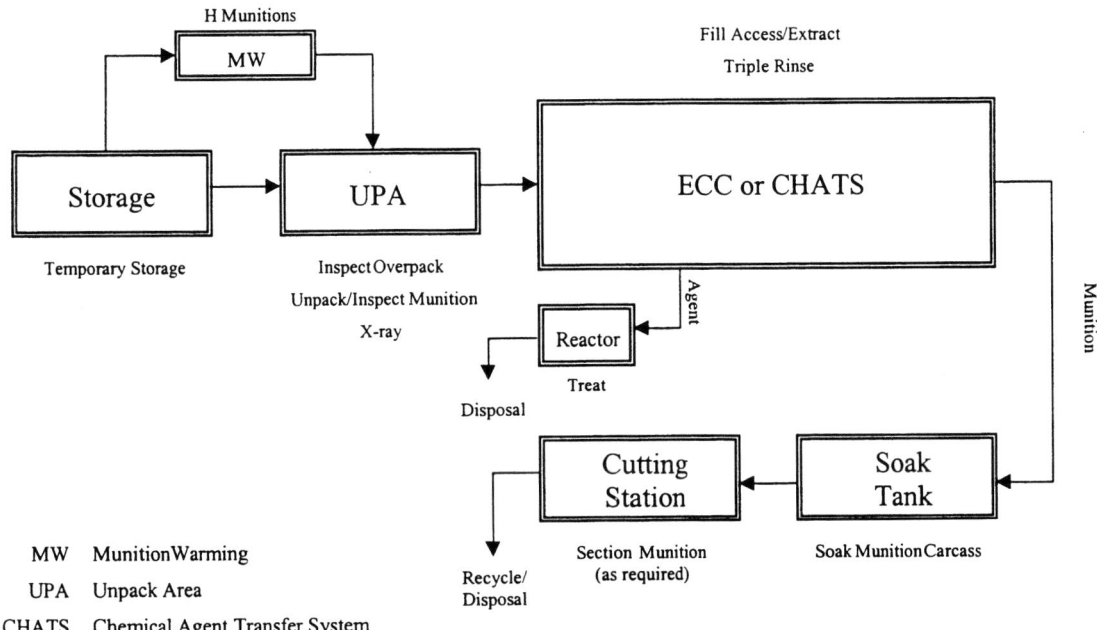

FIGURE D-2 Typical process flow for non-explosively configured munitions at PBNSF. SOURCE: U.S. Army (2000).

tent, crush the bottles, and neutralize the chemical agent. The waste stream produced is sent to the secondary treatment facility within the Pine Bluff facility for final treatment. The RRS provides the majority of the equipment used for this process and is part of the CHATS. The typical process flow for chemical agent detection systems is shown at Figure D-3.

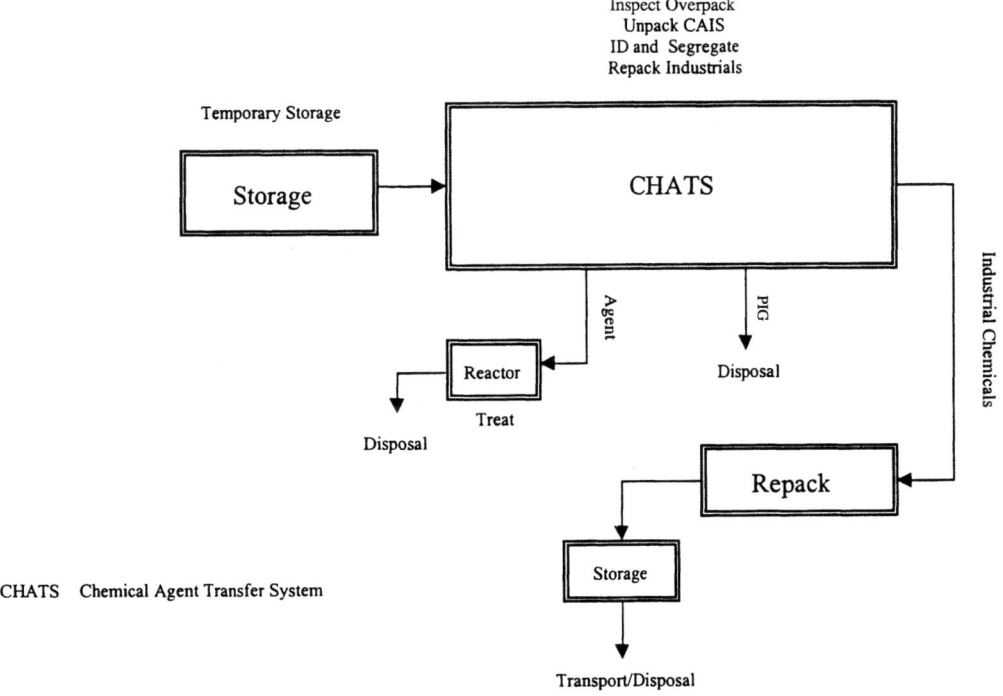

FIGURE D-3 Typical process flow for chemical agent identification sets at PBNSF. SOURCE: U.S. Army (2000).

Process Flow for Chemical Samples

The small amount of CWM in this category will be processed in the PBNSF together with other items having similar chemistry.

REFERENCE

U.S. Army. 2000. Request for Proposal. Pine Bluff Arsenal Fixed Facility (PBA-FF). Solicitation number DAAA09-99-R-0090. October 27. Aberdeen Proving Ground, Md.: Program Manager for Chemical Demilitarization.

E

Mobile Non-Stockpile Systems

This appendix reviews the two mobile non-stockpile facilities proposed for treatment of NSCWM—the Rapid Response System (RRS) and the Explosive Destruction System (EDS).

RAPID RESPONSE SYSTEM

The Rapid Response System (RRS) occupies three trailers: an operations trailer, a support trailer, and a utility trailer. The operations trailer contains the process equipment and instrumentation. The support trailer contains spare equipment and supplies. The utility trailer carries electrical generators to allow the system to operate without commercial or host power when needed. For a more complete description of the RRS equipment and operations, see *Rapid Response System Test Report* (U.S. Army, 2001a).

Chemical agent identification sets (CAIS) were produced in large quantities (approximately 110,000) and in various configurations from 1928 through 1969. The configurations included the following:

- sealed Pyrex tubes or ampoules containing chemical agents (e.g., H/HD, HN, and L) and industrial chemicals (e.g., cyanogen chloride, phosgene, chloropicrin, and chloroform)
- glass bottles containing neat (pure) mustard agent
- "sniff sets"—widemouthed glass jars containing H/HD or L adsorbed on granular charcoal

The reagents employed in the RRS for the destruction of the various agents and industrial chemicals found in CAIS are as follows:

- *Red process.* For use on HN-1, H/HD, and L in chloroform solution, it employs dichlorodimethylhydantoin (DCDMH) in chloroform, t-butyl alcohol, and water.
- *Blue process.* For use on H/HD, it employs DCDMH in chloroform, t-butyl alcohol, and water.
- *Charcoal process.* For use on HN or H/HD adsorbed on charcoal, it employs DCDMH in chloroform.
- *Charcoal-L process.* For use on L adsorbed on charcoal, it employs DCDMH in chloroform, t-butyl alcohol, and water.

The deployment time for the RRS is 2 weeks. The RRS can treat one PIG[1] of CAIS per day. Adding transportation and set-up time, a site with one PIG could be dealt with in 6 weeks. This does not include the paperwork and other actions connected with removing the treatment residues from the site, which could add 2 weeks to the schedule.

Following their recovery, CAIS may be stored for only 90 days without a RCRA permit. After that, they must be moved to a permitted storage facility. The RRS was initially permitted under a RCRA permit by the state of Utah to conduct a test program with both simulants and chemical agents at the Deseret Chemical Depot.

A full-scale prototype was designed and assembled. The state of Utah approved a testing program to qualify the process, and 33 of the 60 sets of CAIS stored at Deseret were destroyed during the program. The operation was carried out

---

[1]A PIG is a metal canister with packing material designed to protect CAIS during transport.

successfully and is documented in detail (Rapid Response System Operations Approval In-Process Review (IPR) Package, January 29, 2001). The treatment goal was to reduce agent concentration to less than 50 ppm. This goal was met, with most residue containers having agent concentrations of less than 1 ppm. The operations were then converted to a production mode, and the remainder of the CAIS, more than 1,200 items, was destroyed. A final report will be issued by the Army contractor, SAIC. Only one negative incident was reported to the committee. When a container was opened, a small quantity of chloromethane was released and passed through all of the filters into the containment building. The source of the chloromethane is unknown. An evaluation of the safety and environmental performance of these operations was published in 2001 (Mitretek, 2001). It notes that poor analytical quality control data resulted in a decision to postpone shipment of the waste drums with decontaminated liquid or solid wastes to TSDFs pending resolution of the issue. These issues were resolved, and the wastes were disposed of at commercial TSDFs (U.S. Army, 2001a, 2001b).

After the Deseret campaign, the RRS was moved to Huntsville, Alabama, for modifications by the contractor. These include increasing the capacity of the electrical power supply system and reconfiguring the system, done in response to a failure of the uninterruptible power supply during operations in Utah. The RRS will then be dispatched to the following locations (in the expected deployment sequence): Fort Richardson, Alaska; Camp Bullis, Texas; Redstone Arsenal, Alabama; and Pine Bluff Arsenal, Arkansas. Operations at Fort Richardson will fall under the existing CERCLA action at that site, without further permitting under RCRA. Drafting of the RCRA permit applications for Camp Bullis, Redstone, and Pine Bluff has started. The Pine Bluff application was to be the template for the others. The Army expects that the order of deployment will be determined by the order in which the permits are approved. Ultimately, the plan is to make Pine Bluff Arsenal the home base of the RRS, where crews will be trained and local CAIS will be destroyed when the RRS has not been dispatched elsewhere.

The RRS is intended for use on multiple items of CAIS. If only one or two items are found at a site, the SCANS system, under development, would be used. The cost of transporting the RRS to a treatment site can be substantial and could have an impact on the Army's disposal decisions. It might be cost effective to use SCANS for a much larger number of items, as previously discussed. A summary of the costs involved with the permitting, transportation, and operation was previously published by this committee (NRC, 1999, p. 79). The costs for the Fort Richardson cleanup were estimated as follows:

Obtaining a RCRA permit	$250,000
Transportation of the equipment from Utah to Alaska	33,000
Transportation of personnel to Alaska	172,000
55 days of operation	
Labor	457,000
Materials and equipment	228,000
Management, engineering, other	250,000
Allocation of construction costs, Other indirect	400,000
Total	$1,790,000

Of interest is the use of SCANS instead of the RRS. SCANS is under development and would be more efficient for dealing with individual CAIS items, although it will not be able to open a PIG to remove individual CAIS items.

EXPLOSIVE DESTRUCTION SYSTEM

The Explosive Destruction System, Phase 1 (EDS-1) is a trailer-mounted mobile system intended to destroy explosively configured chemical warfare munitions that are deemed unsafe to transport or store routinely. The EDS can also be used to destroy limited numbers of chemical munitions, with or without explosive components, when the quantity of these munitions does not require the use of other destruction systems. A detailed description of the EDS and its operation is found in NRC (2001). A schematic view of the EDS-1 is shown in Figure E-1.

The heart of the EDS-1 is a 6.5-cubic-foot (189-liter) explosion containment vessel mounted on a 20-foot-long flatbed trailer. The vessel is fabricated from two 316 stainless steel forgings and is designed to contain detonations of up to one pound (0.45 kg) of TNT equivalent. The explosion containment vessel contains the explosive shock, fragments, and chemical agents during the munition opening process and also serves as a processing vessel for subsequent neutralization of the chemical agent and energetics within the munition.

The EDS-1 has an inside diameter of 51 cm and is designed to handle three common munitions: a 75-mm artillery shell, a 4.2-inch mortar, and a Livens projectile. It has been used to dispose of a 4-inch Stokes mortar and a M-139 bomblet, as well as nonexplosive cylinders.

Once a munition is placed inside the explosion containment vessel and its hinged door is closed and secured with clamps, shaped charges are used to open the munition and detonate any explosives within it. Chemical reagents are then introduced to treat the chemical agent within the munition until agent quantities are reduced to acceptable levels. The liquid neutralent is then treated as a hazardous waste and disposed of.

Prior to placing the munition in the EDS containment vessel, it is placed in a fragment suppression system (FSS) consisting of two steel half-cylinders, one above and one below the munition. The FSS reduces the velocity of small fragments in order to protect the wall of the EDS containment vessel. A steel support connected to the bottom half-

APPENDIX E

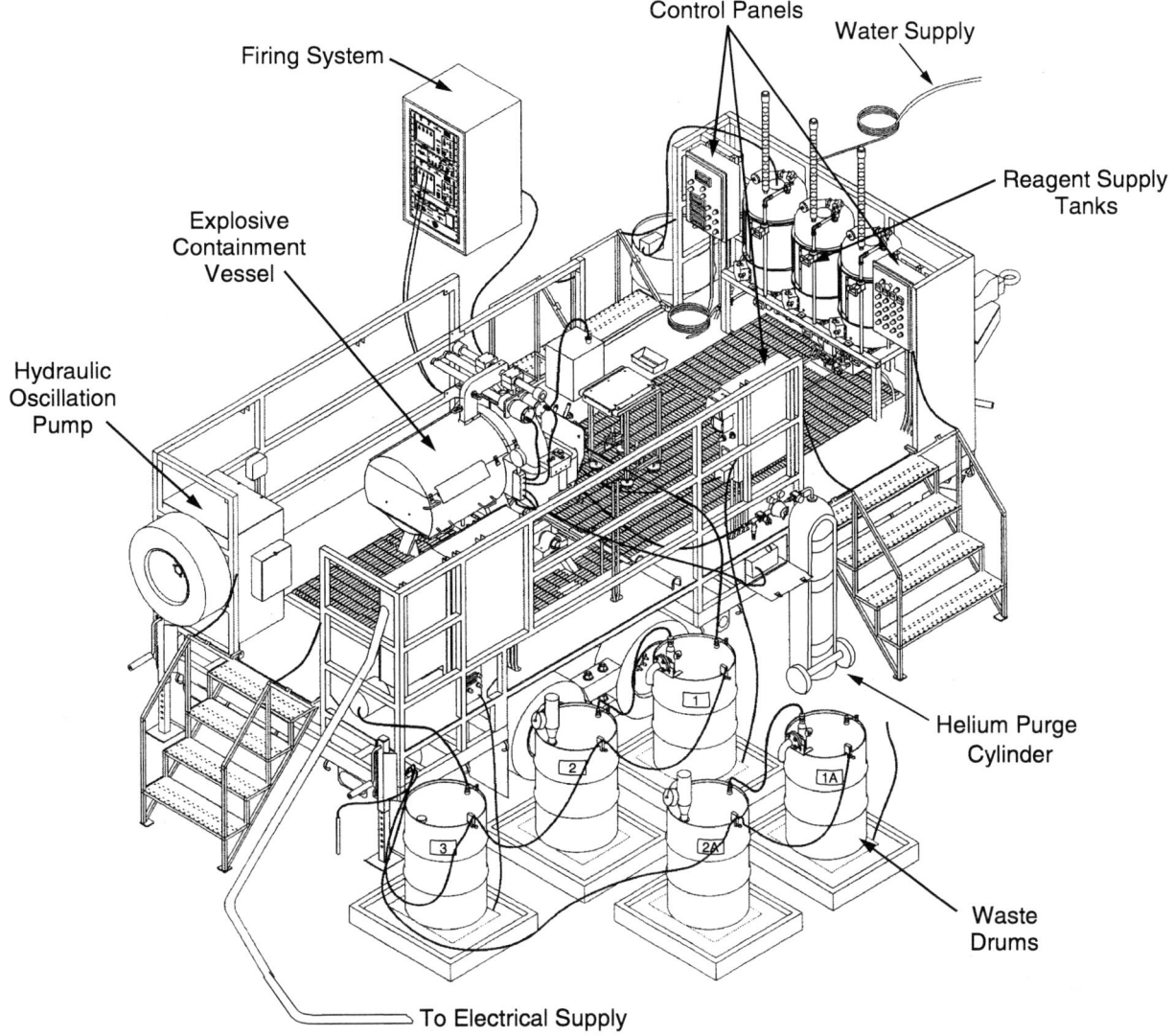

FIGURE E-1 Diagram of the EDS-1 vessel on its trailer. SOURCE: U.S. Army (2001c).

cylinder centers the munition in the FSS and provides shock absorption to protect the lower part of the EDS vessel. Additional protection is provided by a steel block placed beneath the bottom half-cylinder (to protect the lower EDS vessel wall from gas jets from the detonations) and end plates hung on the ends of the FSS cylinder to absorb shock and protect the ends of the EDS vessel. In effect, the munition in the EDS is double wrapped, inside the fragment suppression system as well as in the EDS itself.

The FSS is also used to mount and properly locate the charges used for detonation in the EDS. These are a linear shaped charge that cuts open the munition and exposes its contents for chemical treatment and conical shaped charges that detonate the burster inside the munition. Either one or two conical shaped charges, depending on munition size, are positioned above the munition in the FSS.

After unpacking the munition, it is placed in the FSS. The linear shaped charge is attached to the bottom half-cylinder and the conical shaped charges are attached to the upper half-cylinder. Detonators are then attached to the charges and shorted for safety. The FSS is then placed inside the EDS vessel using a loading table.

Following preparation of the door sealing surface and installation of a new O-ring, the EDS door is closed and a leak test conducted. Detonation then takes place and reagent is pumped into the EDS vessel to treat the chemical fill. Reagents used in the EDS include 22 percent aqueous sodium hydroxide for phosgene, 90 percent MEA/water for nitrogen mustard (HD), and 45 percent MEA/water for the nerve agent GB. Reactions take place at low temperatures and pressures.

Following treatment of the chemical munition or item, the EDS vessel is rinsed, cleaned, and inspected. This includes inspection of the sealing surface and the EDS door as well as replacement of the all-metal seal that contains the detonation and the O-ring seals that provide an airspace used for helium leak detection.

Operational Characteristics

Deployment Protocol

The deployment protocol calls for seven EDS units to be operational by FY 2007, including three Type 1 and four Type 2 units. Two Type 1 units will be based in APG. One will be on standby for deployment as required to handle emergency situations that require timely disposal of unstable munitions. The other will be used at APG to handle remediation activities at Lauderick Creek. The third Type 1 EDS will be based at Pine Bluff for destroying warfare material recovered at that location.

The EDS developer, Sandia National Laboratories, is also designing and fabricating an EDS Phase 2 (EDS-2). The EDS-2 vessel will be capable of repeated use cycles at 3 pounds TNT equivalent and occasional uses of 5 pounds of TNT, should such a need arise. The frequency of allowable use above 3 pounds has yet to be defined. The EDS-2 vessel will be tested at Livermore Laboratory at more than 5 pounds of TNT equivalent for rating purposes. This larger version of the EDS will be able to dispose of munitions as large as 155-mm projectiles. The Army plans to have four EDS-2 units: two for operations and the other two as replacement units. Developmental testing of the first EDS-2 is planned to take place in Porton Down, United Kingdom, in 2003, followed by operational testing at APG, also in 2003. This unit is to be operational by FY 2004-2005 and available to support emergency response and chemical samples disposal. The purchase of an additional three Type 2 units is planned for FY 2006-2007 as required.

Throughput

Table E-1 presents some approximate EDS processing times for different agent fills. The data were derived from test and deployment results using the EDS system. In addition, there can be 3 hours of heat-up time for hot water rinses.

The best performance at RMA on sarin was destruction of one munition every 2 days. This would roughly be 8-9 hours of processing time on the first day, with the second day devoted to preparing (replacing door seal, charging reagent tanks, etc.) the EDS unit for the next munition.

Secondary Waste Streams

Based on current and expected operation of the EDS, three types of liquid wastes are generated:

1. neutralent, consisting of the initial treatment of agent with active reagent (e.g., MEA) and any subsequent chamber washes with chemical reagent (if used)
2. rinsate, consisting of additional agent treatment with water and chamber washes with water after opening the EDS
3. cleaning solution, consisting of washes (water/detergent) that are made between processing of each munition, and final washes (e.g., water/acetic acid) made after completing a munitions campaign

The expected source and collection of these wastes are presented in Table 2-2.

TABLE E-1 Approximate EDS Processing Time, by Agent

Agent Fill	Treatment Time (hr)	Rinse Times (hr)	Total Contact Time (hr)
Phosgene	1-1.5	0.17	1.2-1.7
Mustard	4-4.5	2.25	6.3-6.8
Sarin	4.25-4.5	1	5.3-5.5

SOURCE: NRC (2001).

REFERENCES

Mitretek. 2001. D. Tripler, K. Raghuveer, R. Rhoads, A. Amr, and J. Miller. Independent Evaluation of the Rapid Response System, Safety and Environmental Performance, Mitretek Technical Report MTR 2001-22. Aberdeen Proving Ground, Md.: Program Manager for Chemical Demilitarization.

NRC (National Research Council). 1999. Disposal of Chemical Agent Identification Sets. Committee on Review and Evaluation of the Army Non-Stockpile Chemical Materiel Disposal Program. Washington, D.C.: National Academy Press.

NRC. 2001. Evaluation of Alternative Technologies for Disposal of Liquid Wastes from the Explosive Destruction System. Committee on Review and Evaluation of the Army Non-Stockpile Chemical Materiel Disposal Program. Washington, D.C.: National Academy Press.

U.S. Army. 2001a. Rapid Response System Test Report. Final. October. Aberdeen Proving Ground, Md.: Program Manager for Chemical Demilitarization.

U.S. Army. 2001b. Non-Stockpile Systems Program, Rapid Response System (RRS) After Action Report. Contract No. DAAA-09-95-D-0001, Task 0019. November. Prepared by Teledyne Brown Engineering.

U.S. Army. 2001c. Emergency Destruction Plan for M139 Bomblets Filled with GB at Rocky Mountain Arsenal, Colorado, using the Explosive Destruction System (EDS). Product Manager for Non-Stockpile Chemical Materiel, Aberdeen, Md. Final Revision 3, January 2001.

F

Regulatory Background

This appendix provides background information on the statutory and regulatory programs that impact RAP for NSCWM activities. It concludes with a discussion of available RAP mechanisms for treatment of NSCWM.

THE RESOURCE CONSERVATION AND RECOVERY ACT

RCRA was intended by Congress to be a state-implemented program. Under this program, the U.S. Environmental Protection Agency (EPA) was charged with developing regulations that would define wastes as hazardous and establish a cradle-to-grave system for managing them. States would then adopt these regulations and seek authorization from EPA to implement the RCRA program within their respective boundaries. EPA implements the program within states that choose not to adopt and implement the hazardous waste program.[1] Those states that choose to adopt EPA's regulations may decide to seek authorization from EPA to implement the entire RCRA program or a portion of it. In some states, therefore, facilities are regulated by the state for some activities and by the EPA for other activities. Additionally, when they adopt EPA's regulations, the states cannot be less stringent than the federal program, but they can develop regulations for implementation within their state that are more stringent or broader in scope. For example, states have the authority under RCRA to regulate additional waste streams as hazardous within their boundaries that EPA does not regulate.

Under the RCRA program, wastes may be designated hazardous waste under two separate and distinct mechanisms. First, wastes may be listed as hazardous waste. Six of the eight stockpile states—Colorado, Indiana, Kentucky, Maryland, Oregon, and Utah[2]—list military chemical agents as hazardous waste. These wastes are not listed hazardous waste under EPA regulations or in other states.

Under the hazardous waste listings, wastes may be listed for various reasons. For example, they may be listed because they are toxic or acutely toxic. RCRA requirements for management of acutely toxic wastes are more restrictive than its requirements associated with other types of hazardous waste. For example, RCRA places more stringent restrictions on the amount of acutely toxic waste (as compared with other types of hazardous waste) that may be stored at any one time.

In addition, pursuant to the RCRA mixture and derived-from rules—40 CFR §261.3(a)(2)(iv) and 40 CFR §261.3(c)(2)(i), respectively—wastes that are mixed with listed hazardous waste or that are derived from the treatment, storage, or disposal of listed hazardous waste (e.g., treatment residues) are themselves regulated as listed hazardous waste. RCRA, however, provides an exclusion mechanism, known as a delisting petition, that generators may use to demonstrate that a listed hazardous waste, including mixture and derived-from waste, is not hazardous (40 CFR §260.22). Delistings are risk-based exclusions that

---

[1] All the stockpile states are authorized by EPA to operate their own RCRA programs in lieu of EPA.

[2] Interestingly, each of these states lists chemical agents (and in some cases, associated wastes) differently.

apply to specific wastes generated at specific facilities. Many facilities have pursued delistings for mixture and derived-from wastes, because in many cases these wastes no longer contain hazardous constituents in significant concentrations.

Historically, however, the delisting process has been long and arduous, often taking months or years and many thousands of dollars to achieve. EPA has attempted to develop a rule that would provide a more generic and self-implementing process for relieving wastes that could qualify for delisting from hazardous waste regulation. Known as the Hazardous Waste Identification Rule (HWIR), it would provide a risk-based mechanism for relieving low-risk wastes from hazardous waste control. Unlike delistings, however, HWIR would apply to all wastes at all locations. Using a complex multimedia model, the HWIR would provide concentrations of hazardous constituents below which a waste would no longer be considered hazardous. The HWIR rule has been very controversial, however. On November 19, 1999, EPA released its proposed HWIR (64 FR 63382). The proposed HWIR constituted a reproposal of a 1995 proposed rule (60 FR 66344, November 13, 1995), which was highly criticized by both state regulators and the regulated community. While EPA intends to finalize the HWIR rule, the final form of the rule and the date of promulgation remain uncertain.[3]

The second mechanism by which wastes may be designated hazardous waste is by the federal RCRA hazardous waste characteristics. These characteristics include ignitability, corrosivity, reactivity, and toxicity (40 CFR §§261.21-261.24), as follows:

Ignitability (40 CFR §261.21)—An ignitable waste is a waste that (1) is a liquid, other than an aqueous solution containing less than 24 percent alcohol by volume, and has a flash point less than 60 °C or 140 °F as determined by various test methods; (2) is not a liquid and is capable, under standard temperature and pressure, of causing fire through friction, absorption of moisture, or other spontaneous chemical changes and, when ignited, burns so vigorously and persistently that it creates a hazard; (3) is an ignitable compressed gas; or (4) is an oxidizer as defined in U.S. Department of Transportation (DOT) regulations.

Corrosivity (40 CFR §261.22)—A corrosive waste is a waste that (1) is aqueous and has a pH less than or equal to 2 or greater than or equal to 12.5; or (2) is a liquid and corrodes steel (SAE 1020) at a rate greater than 6.35 mm (0.250 inch) per year at a test temperature of 55 °C (130 °F).

Reactivity (40 CFR §261.23)—A reactive waste is a waste that (1) is normally unstable and readily undergoes violent change without detonating; (2) reacts violently with water; (3) forms potentially explosive mixtures with water; (4) when mixed with water or, when it is a cyanide- or sulfide-bearing waste that, when exposed to pH conditions between 2.0 and 12.5, can generate toxic gases, vapors, or fumes in a quantity sufficient to present a danger to human health or the environment; (5) is capable of detonation or explosive reaction if it is subjected to a strong initiating source or if heated under confinement; (6) is readily capable of detonation or explosive decomposition or reaction at standard temperature and pressure; or (7) is a forbidden explosive as defined in DOT regulations.

Toxicity (40 CFR §261.24)—A toxic waste is a waste that contains concentrations of certain listed contaminants above established thresholds when tested using the Toxicity Characteristic Leaching Procedure, a leaching test. Contaminants include both organic (e.g., pesticides, chlorinated organics) and inorganic contaminants (e.g., heavy metals, including As, Ba, Cd, Cr, Hg, Pb, Se and Ag).

Unlike listed hazardous waste, wastes that exhibit one or more of the RCRA characteristics are not subject to the RCRA mixture or derived-from rules. In addition, a waste may be a listed hazardous waste and also exhibit one or more hazardous waste characteristics.

Once wastes are designated hazardous waste, by either listing or characteristic (or both), generators and RCRA treatment, storage, and disposal facilities (TSDFs) are subject to a wide range of regulations. Of particular note for this report is the requirement to obtain a permit for treatment, storage, or disposal operations. In some cases, however, RCRA provides alternative mechanisms for obtaining approval of the regulator, and several of these alternative mechanisms are particularly well suited for NSCWM treatment operations. RAP mechanisms for NSCWM are reviewed later in this appendix.

Also of note for this report are the RCRA Land Disposal Restrictions (LDRs).[4] The LDR program was mandated by the RCRA Hazardous and Solid Waste Amendments of 1984. In essence, LDRs are treatment standards for listed and characteristic hazardous waste that must be achieved prior to land disposal. Treatment standards under the LDR program are established on the basis of the best demonstrated available technology and are therefore technology-based (as opposed to risk-based). Although LDR standards are technology-based, EPA has proposed, as part of HWIR, to cap LDR treatment standards with the HWIR risk-based levels. In this manner, treatment would not be required below those levels necessary to minimize risk to human health or the environment.

LDRs would apply to non-stockpile wastes (or stockpile wastes) only if they exhibit one or more of the RCRA characteristics. Because EPA has not listed any agent waste as hazardous waste, LDR treatment standards have not been established specifically for these wastes. To date, no state that lists agent waste as hazardous has established LDRs or other regulatory-based treatment requirements for these wastes.[5] Because most states operate their own RCRA regu-

---

[3]EPA has finalized some provisions of the original HWIR rule. However, that aspect of the rule that would establish risk-based exemption levels has not been finalized.

[4]LDRs are established in 40 CFR §268.

[5]Some states, such as Kentucky, have established a generic treatment requirement for agent wastes (e.g., 99.9999 percent destruction efficiency), but specific treatment standards have not been established.

latory program in lieu of EPA, and considering that most of the stockpile states have listed chemical agents as hazardous waste, the remainder of this appendix assumes that NSCWM waste would be managed under a state-implemented program. It should be recognized, however, that in some states, EPA implements the RCRA program.[6]

THE COMPREHENSIVE ENVIRONMENTAL RESPONSE, COMPENSATION AND LIABILITY ACT

The Comprehensive Environmental Response, Compensation and Liability Act (CERCLA) is a federal law governing the cleanup of releases of hazardous substances, pollutants or contaminants from uncontrolled hazardous waste sites (40 CFR 300). CERCLA, as implemented by the National Contingency Plan,[7] applies to the private sector as well as to government agencies, including the Department of Defense (DOD) and the Army. Unlike RCRA, which is intended to be a state-implemented program, CERCLA is implemented by the federal government. Whereas EPA implements the program for private sector entities and has oversight over federal government cleanups, the DOD is designated the lead agency for the Army with respect to CERCLA actions. The states and the public nevertheless have a significant role in CERCLA implementation.

There are basically two separate cleanup processes prescribed by CERCLA and the National Contingency Plan. Removal actions are intended to address emergency situations or time-critical concerns. While they can entail permanent remedies, in general, removal actions are intended as a temporary remedy to prevent, minimize, or mitigate a release of hazardous substances. Remedial actions, on the other hand, entail a more thorough evaluation of site conditions and risks and result in permanent remedies.

For federal agencies, including Army installations, CERCLA applies to essentially two types of hazardous waste sites: sites listed on the National Priorities List (NPL) and non-NPL sites. Sites listed on the NPL (published in the *Federal Register*) are those that have indicated significant risk as determined through application of a numerical ranking system (i.e., the Hazard Ranking System). Non-NPL sites may nevertheless present significant risk and are therefore also addressed under CERCLA. Either or both removal and remedial actions may be taken at NPL and non-NPL sites.

Actions taken pursuant to CERCLA removal authority may be taken at any location, even at RCRA facilities, by the lead federal agency (in this case the DOD) without prior regulatory approval. While federal agencies are required to coordinate these types of CERCLA actions with state regulators, there are no federal or state approvals required. Because DOD is the lead agency for Army installations, DOD can make CERCLA removal decisions without the approval of regulatory authorities, even at NPL sites. Decisions under the remedial program, however, require the concurrence of the EPA and the states. For removal actions, CERCLA nevertheless requires the DOD to coordinate removal decisions and related actions with federal and state regulators. At least for some types of military ranges (e.g., training ranges), such as those closed under the Base Realignment and Closure Program, DOD and EPA have indicated a preference for emergency actions involving ordnance and explosives (which can include NSCWM) to be processed under CERCLA authorities (DOE, 1997; EPA, 2000). Such actions can include remediation measures as well as implementation of treatment technologies such as the RRS and EDS.

The U.S. EPA and DOE have developed numerous CERCLA guidance documents that may be consulted for more information on CERCLA removals and remedial actions.[8]

THE CHEMICAL WEAPONS CONVENTION

The CWC also affects management of non-stockpile items and associated wastes. As indicated previously, about 85 percent of all recovered CWM in the United States is stored at Pine Bluff Arsenal in Arkansas; smaller quantities are stored at Aberdeen Proving Ground in Maryland and Dugway Proving Ground in Utah. According to the CWC, these recovered items must be destroyed by April 2007. NSCWM buried before January 1997 is excluded from treaty requirements as long as it remains buried (U.S. Army, 2001). NSCWM removed from burial sites, however, must be identified and declared under the CWC.

The CWC addresses the destruction of the agents, which are defined under the CWC as Schedule 1 compounds, but it also addresses agent precursors, known as CWC Schedule 2 compounds. Schedule 2 compounds can be present in some types of agent treatment residues, such as RRS and EDS neutralent. These compounds are not considered hazardous constituents under the federal RCRA program; however, some states have shown concern for some of these compounds in their regulatory programs. Like the chemical agents, there are no federal or state RCRA standards for their treatment. Further, applicability of the CWC to wastes that may contain low concentrations of Schedule 2 compounds

---

[6]RCRA state authorization status may be viewed for each state at the following EPA Website: <http://www.epa.gov/oswer/hazwaste/state/index.htm#adoption>.

[7]Formally, the National Oil and Hazardous Substances Pollution Contingency Plan (40 CFR 300).

[8]These guidance documents may be obtained through EPA and DOE Web sites <http://www.epa.gov/> and <http://www.doe.gov/>, respectively.

(such as EDS or RRS secondary wastes) is unclear. If the CWC places restrictions on wastes containing small amounts of Schedule 2 compounds, off-site treatment at commercial TSDFs may be more difficult.

Overall, the CWC requires that stockpile and non-stockpile munitions be destroyed in accordance with a very aggressive schedule in order to meet the established deadlines. Because RCRA permits for some non-stockpile operations may take several years or more to achieve, when RCRA permits are required, the Army clearly faces a daunting challenge in meeting CWC schedule requirements.

ARMY REGULATION AR 50-6

On June 26, 2001, the U.S. Army revised Army Regulation (AR) 50-6 for chemical surety. The surety program outlines a system of reliability, safety, and security control measures designed to protect the local population, workers, and the environment by ensuring that only personnel who meet the highest standards of reliability conduct chemical agent operations, that chemical agent operations are conducted safely, and that chemical agents are secure. AR 50-6 applies to all Army programs that involve chemical agents, including the stockpile and non-stockpile programs.

The primary component of AR 50-6 that affects the non-stockpile program is that pertaining to recovered chemical warfare material (RCWM). These requirements are described in Chapter 12 of AR 50-6. The specific definition of RCWM is as follows:

> Chemical agent material and/or associated equipment and surrounding contaminated media discovered either by chance or during deliberate real estate recovery/restoration operations that was used for its intended purpose or previously disposed of as waste. RCWM will be classified based on the requirements of 40 CFR 266 Subpart M (EPA Military Munitions Rule). RCWM does not fall within the scope of the Army Chemical Surety Program except as detailed in Chapter 12.

Chapter 12 outlines numerous requirements for RCWM. In general, these provisions are not as comprehensive as those associated with surety material, but they are nevertheless substantial. In addition, although AR 50-6 is focused on surety measures, it addresses other areas that are important in the committee's analyses. For example, the AR indicates that RCWM found buried is required to be managed in compliance with environmental laws and regulations, including CERCLA and RCRA, and must be classified (regarding whether it is hazardous waste or not) per the RCRA Military Munitions Rule.[9] Also, the AR specifies that deliberate unearthing of suspect RCWM will not begin until all required plans and approvals are obtained for transportation and storage or treatment.

AR 50-6 also defines requirements for management of RCWM. The regulation specifies that emergency on-site destruction of chemical munitions may be considered an option to reduce risk and that non-emergency on-site destruction is subject to CERCLA or RCRA. Also, AR 50-6 indicates that soil suspected of contamination by chemical agents or industrial chemicals is presumed hazardous until confirmed otherwise by laboratory analysis and is required to be managed in accordance with environmental laws and regulations.

The AR also addresses requirements for transportation of RCWM. In general, RCWM may be transported to an approved location if on-site treatment or disposal cannot be accomplished. Once transported to an installation with an active surety mission, RCWM is afforded the same safety and security measures as chemical surety material. The regulation states, though, that this is not to reclassify RCWM as surety material.

One of the more significant new provisions of the revised AR is the classification of CAIS as RCWM. This decision triggers many requirements, including notification of states and Congress prior to transportation across state lines. As discussed in Chapter 4 of this report, the committee believes that it is not necessary to handle CAIS as RCWM.

SUMMARY OF RAP MECHANISMS AND PROCESSES

The following paragraphs provide a summary of the processes that could be used for RAP of non-stockpile treatment technologies. In most cases, RAP for treatment of NSCWM is a responsibility of state regulators; however, expeditious RAP requires coordination and cooperation between the state regulators and the Army, as well as the public. Although EPA does not specifically regulate chemical agents under the RCRA program, it does play a consulting role for the states.

RCRA Permitting Process

40 CFR §270, as adopted and implemented by the states, provides requirements for facilities seeking permits for management of hazardous waste and requirements for regulators who issue these permits. The RCRA permitting rules are quite complex and are the subject of a number of guidance documents that describe the permitting process (EPA, 1984; DOE, 1997). Relevant points are the following:

• RCRA permits pertain primarily to the units used to manage the waste (e.g., landfills, incinerators) and are tailored to address management of particular types of wastes in these units.

---

[9] Among other things, the RCRA military munitions rule (62 FR 6622, February 12, 1997) clarifies applicability of RCRA to waste military munitions, including chemical munitions.

- RCRA permits are initiated through an application process where the facility submits a permit application to the regulator. Following submittal of the permit application, negotiations between the regulator and the permittee take place until permit conditions are agreed.
- The permit application and negotiation process, even for simple storage permits, typically takes over a year. Permits for landfills or incinerators typically take 3 years or more to permit.
- There are requirements for public involvement at several points along the permitting path.
- The RCRA regulations have many gray areas where interpretation can be difficult and contentious.
- Changes in facility operations during the permit life are facilitated through permit modifications.
- RCRA corrective action[10] decisions for cleanup of releases from solid waste management units, some of which could contain buried NSCWM, could also be facilitated as a permit modification. At some facilities, and in particular those that have not yet received an operating permit, RCRA corrective action requirements may be issued through a RCRA order, described later in this appendix.

In most cases, conventional RCRA permits pertain to moderate- or long-term operations involving fixed or semipermanent facilities. PMCD's incinerator operations, for example, are permitted through this mechanism. Some NSCWM operations—and particularly longer-term ones such as MAPS and PBNSF— will be permitted as a RCRA treatment, storage, and disposal facility (TSDF). Initial operation of these types of facilities, however, could be approved through a different mechanism, such as a RCRA Research, Development, and Demonstration (RD&D) permit. In fact, operation of MAPS was initiated as a RCRA RD&D permit, but it is to transition into a regular RCRA permit once operations become routine. Operation of PBNSF is likely to follow a similar path.

The RRS was permitted in Utah (Deseret Chemical Depot) as a RCRA TSDF. The MMD-1 was permitted in Utah (Dugway Proving Ground) under a RCRA RD&D permit (discussed later in this appendix).[11] The committee expects that some NSCWM operations will not warrant the establishment of conventional RCRA permits. For example, the RRS or EDS (or similar treatment systems) could be approved under RCRA or CERCLA emergency actions (discussed further below). However, treatment of secondary and other wastes, such as RRS and EDS neutralents and rinsates, could be managed at on-site or off-site TSDFs. If RRS and EDS neutralent and rinsates are to be managed at existing facilities (including stockpile facilities)[12] under TSDF permits, a permit modification may be required.

Emergency Responses

In many cases, the RRS or EDS systems (or similar treatment systems) may be used for emergency destruction operations for stable items—those that can be safely moved. Unstable non-stockpile items—those too sensitive to move—may be treated using alternative technologies being developed by the Army. While open detonation has been used to destroy these types of items in the past, the Army is developing a tent-and-foam technology that, if successful, would permit detonation while minimizing subsequent dispersion of detonation by-products, including potential unreacted agent.

Both stable and unstable NSCWM could be addressed under the emergency response provisions of the RCRA or CERCLA programs. True emergency events were never intended to be addressed under the conventional RCRA permit, which as indicated above, may take years to obtain. RCRA emergency permits (40 CFR §270.61) or CERCLA emergency removal actions (40 CFR §300.415) may be employed in these cases. In some situations, however, even RCRA's emergency permit provisions are inapplicable. EPA specifically addressed munition emergencies in its Military Munitions Rule (MMR) (EPA, 1997):

> Today's rule clarifies that EPA considers immediate or time-critical responses to explosives or munitions emergency responses to be an immediate response to a discharge or imminent and substantial threat of a discharge of a hazardous waste under 40 CFR Secs. 264.1(g)(8), 265.1(c)(11), and 270.1(c)(3). Such responses are, therefore, exempt from RCRA permitting, and other substantive requirements, including emergency permits. . . . If an immediate response, however, is clearly not necessary to address the situation, and a response can be delayed without compromising safety or increasing the risks posed to life, property, health, or the environment, the responding personnel, if time permits, should consult with the regulatory agency regarding the appropriate course of action (e.g., whether or not to seek a RCRA emergency permit under Sec. 270.61, or regular facility permit under 40 CFR Part 270). Situations where an immediate response is needed would include instances where the public or property is potentially threatened by an explosion. Situations where an immediate response is clearly not necessary would include instances where the public or property are not threatened by a potential explosion (e.g., in remote areas such as some former ranges or where

---

[10]Mandated by the Hazardous and Solid Waste Amendments to RCRA in 1984, RCRA corrective action requires cleanup of releases of hazardous waste or hazardous constituents from solid waste management units at RCRA facilities (RCRA Section 3004 (u) and (v)). Similar to the CERCLA program, RCRA corrective action pertains specifically to cleanup at RCRA facilities.

[11]Note that the MMD-1 was used to treat phosgene. The Army decided to discontinue its use prior to treatment of agent.

[12]Permit modification may be especially warranted at stockpile facilities, because in most cases, these permits prohibit the Army from accepting wastes generated from off-site locations, which would include NSCWM or NSCWM secondary wastes.

immediate action is not necessary to prevent explosion or exposure). In these cases, there is time to consult with the EPA or state regulatory agency on how to proceed.

Further, in the final MMR, EPA evaluated the DOD statutory requirements and standing operating procedures and found the emergency response procedures sufficiently protective for chemical munitions responses. For example, transport and destruction of lethal chemical agents are regulated by 50 U.S.C. 1512 and 1512a, requiring special approvals by the Secretary of Defense and the Secretary of Health and Human Services prior to transport or destruction. Congress and affected state governors must also be notified prior to any such destruction or transportation. The standards for emergency responses in the final MMR, including the exemption for immediate responses and the requirements for emergency permits, would apply equally to conventional and chemical munitions.

As indicated above, another RAP option for operation of the RRS or EDS device (in lieu of the RCRA emergency permit) is use of CERCLA emergency removal authority. This RAP mechanism is similar to the RCRA emergency permit in nature and scope, and in many situations, either regulatory authority may be applied. The primary difference between RCRA emergency permits and CERCLA removal authority is that actions taken pursuant to a RCRA emergency permit are conducted under state authority and under state oversight and reporting. In most situations, either RCRA emergency permits or CERCLA emergency removal authority may be used for RAP in situations that warrant use of the EDS or the RRS. However, if an emergency RAP mechanism is employed for operation of the RRS or EDS, the RAP documentation may or may not include requirements for management of secondary wastes. Such wastes may be deferred to on-site waste management units (if the response is conducted at an existing installation), or to treatment at an off-site TSDF (assuming the secondary or other waste is defined as hazardous). The emergency that warranted operation of a primary treatment technology such as the RRS or EDS should not extend to treatment of secondary wastes, such as RRS or EDS neutralent and rinsates. The committee believes that once the primary threat of release of the chemical agent is over, the emergency situation is over as well.

RCRA Orders

RCRA orders constitute a different RAP mechanism that could be used for NSCWM operations. There are several different types of orders that could be applied, including a RCRA §7003 imminent and substantial endangerment order, a RCRA §3008(h) corrective action order, or the state equivalents to these orders. The §7003 order would typically be used in emergency situations. These orders are issued by the regulator when an imminent emergency is perceived.

While a RCRA emergency permit could also be used in these situations, use of the §7003 order provides additional latitude to the regulator to take immediate action. Operation of the EDS to destroy the GB bomblets at RMA was processed under a state equivalent to the RCRA §7003 order.[13]

The §3008(h) order (or state equivalent) could also be applied for recovery and treatment of NSCWM. These types of orders are typically issued by regulators to facilities seeking RCRA permits that have solid waste management units (which could include NSCWM) which have released hazardous waste or constituents into the environment. These facilities are subject to RCRA corrective action requirements. In this case, the §3008(h) order could be exercised. This type of order would typically be used for non-emergency cleanup activities and may include requirements for remediation, including recovery and treatment of NSCWM.

Other RCRA Mechanisms

Other RAP mechanisms include the RCRA treatability study (40 CFR §261.4(e) and (f)) and the RD&D Permit (40 CFR §270.65). As indicated previously, the PMNSCM used the RCRA RD&D permit for demonstration of the MMD-1 at Deseret Chemical Depot. In addition, construction and initial operation of MAPS are being conducted under a RCRA RD&D permit, and after operations become routine, a full RCRA operating permit will be pursued. Permitting for PBNSF is likely to follow a similar path.

These types of RAP mechanisms are used to demonstrate new technologies or to demonstrate that existing technologies can be used to treat new or different types of wastes. These RAP mechanisms usually entail waste treatment and allow an expedited mechanism for RD&D. Both of these RAP mechanisms are limited to treatment of specified waste amounts over a specified period of time.

Other CERCLA Mechanisms

Other CERCLA mechanisms for RAP may also be employed, depending on the specific nature of the activity. These include nonemergency removal actions, as well as remedial actions. Since some EDS and RRS operations will be emergencies, it is unlikely that CERCLA remedial actions would be considered. Unlike CERCLA emergency removal actions, remedial actions under CERCLA require concurrence of the federal and state regulator as part of the CERCLA Record of Decision.

---

[13]It is noted that the RMA responded to the incident using its emergency removal authority under CERCLA. In this case, both RCRA and CERCLA RAP mechanisms were used.

REFERENCES

DOE (Department of Energy). 1997. RCRA Permitting Guide for Hazardous & Radioactive Mixed Waste Management Facilities. DOE/EH[RCRA]9705. U.S. Department of Energy. Washington, D.C.

EPA (Environmental Protection Agency). 1984. Permit Writers Guide. U.S. Environmental Protection Agency. Washington, D.C.: Environmental Protection Agency.

EPA. 1997. Military Munitions Rule. 62 FR 6621, February 12. Washington, D.C.: Environmental Protection Agency.

EPA. 2000. Potential Applicability of Assembled Chemical Weapons Assessment Technologies to RCRA Waste Streams and Contaminated Media. EPA 542-R-00-004, August. Available at <www.epa.gov/tio/clu-in.org>.

U.S. Army. 2001. Final Programmatic Environmental Impact Statement. Vol. 1. Aberdeen Proving Ground, Md.: Program Manager for Chemical Demilitarization.

G

Transportation of Chemical Warfare Materiel

There have long been significant concerns with the transportation of chemical warfare materiel. These concerns have been one of the driving forces behind the development of transportable treatment systems by the Army. This section presents an overview of transportation alternatives and related issues for the non-stockpile program.

TRANSPORTING NON-STOCKPILE ITEMS TO TREATMENT OR STORAGE FACILITIES

Transporting Agent/Untreated Non-Stockpile Items

One option is to transport untreated non-stockpile CWM to the closest available mobile treatment system, regional treatment facility, or other fixed site treatment facility, including stockpile treatment facilities. This would prevent the need to make the substantial investment of time and resources to move transportable facilities to every CWM location, especially for small quantities of chemical agent identification sets (CAIS) or other NSCWM that are discovered in remote locations. Regardless of the intended destination, such transportation would, however, require compliance with the stringent laws and regulations applicable to any chemical warfare materiel.

Legal Requirements Specific to CWM

There are serious restrictions to moving any quantity of CWM under P.L. 91-121 and P.L. 103-337. As far back as 1969 (P.L. 91-121), Congress placed substantial restrictions on transportation of CWM, including requiring advance notification and coordination of shipments with the Department of Health and Human Services (HHS) and Congress, unless under emergency conditions. In 1995 Congress (P.L. 103-337) placed restrictions on moving non-stockpile CWM out of any state except to the closest permitted CWM storage facility, and then only under very strict conditions. Public concern with transportation has effectively foreclosed even this option except under extraordinary situations.

Specific requirements for transporting CWM include the following:

- DOD must first make a determination that the proposed transportation is necessary in the interests of national security.
- DOD must present its transportation plan to the Secretary of HHS and then implement any additional measures recommended by HHS.
- DOD must provide a notification at least 10 days before each shipment to Congress and to the governors of each state through which the CWM will be transported.
- The President may make a determination that the HHS recommendations would have the effect of preventing transportation and may override the HHS recommendations based on national security.
- P.L. 91-441 (1970) provides an exemption from the above review and reporting requirements if the CWM is determined to pose an immediate threat to health and safety.

THE ARMY'S TRANSPORTATION PLANS

The Army plans for complying with legal requirements and transporting CWM are outlined in its Final Programmatic Environmental Impact Statement (U.S. Army, 2001).

It includes two potential modes of transportation for off-site shipments of CWM: (1) military aircraft and (2) truck. Choice of mode is based on distance to be transported and the quantity of CWM. Military aircraft is preferred, including both helicopter and fixed-wing aircraft. Helicopters are used if the distance is within the range of the helicopter without refueling or to move the CWM from the location where it is found to a military airfield for shipment by fixed-wing aircraft.

Trucks could be used for transportation from the recovery site to a military airfield instead of helicopters. The Army plans also allow for trucks to be used for the entire distance from recovery site to the eventual treatment or storage facility, but only if air transportation to the site is not possible or practical.

Other components of the Army's transportation plans include the following:

- Packaging—the Army will transport in the same container as stored unless the CWM is in an unapproved container, in which case it will be repackaged into a DOT-approved container or overpack.
- The overpacked non-stockpile CWM will be checked with air monitors for leaking or contamination before being placed on the aircraft or truck.
- Three escort vehicles, each staffed with two military personnel, will accompany the truck movements.
- Military emergency response personnel will be placed on alert during each shipment and be ready to implement the emergency response plan for each shipment.
- A route-specific transportation plan will be prepared for each shipment, including plans for packaging, escorts, notifications, monitoring, mode of transport, emergency response, and routing.
- The transportation plan will be submitted to the HHS, Congress, and applicable states within the 10-day legal notice time and will be preceded by a site-specific and route-specific hazard analysis.

GENERAL HAZARDOUS MATERIALS TRANSPORTATION REGULATORY REQUIREMENTS

In addition to the plans described above, the Army must comply with applicable federal and state regulations for the transport of hazardous materials. The primary governing body of regulations is the Hazardous Materials Transportation Act (HMTA) administered by DOT and the Resource Conservation and Recovery Act (RCRA) administered by EPA. Any potentially hazardous material must be tested to determine whether it meets DOT's criteria (explosive, corrosive, flammable, radioactive, etc.) for a hazardous material. Once a material is determined to fall within one or more of DOT's hazard classes, 49 CFR then prescribes shipping paper, packaging, marking, labeling, placarding and other operational requirements that apply to that material. If the material is also a waste material intended for treatment or disposal it is subject to RCRA transport requirements. Fortunately, EPA and DOT have a joint agreement that helps to avoid regulatory duplication. EPA has ruled that individuals who generate or transport hazardous waste and who have complied with all applicable DOT hazardous materials regulations, are considered compliant with EPA hazardous waste transport regulations as long as they have obtained an EPA identification number and follow EPA manifest requirements.

Most states have adopted the federal HMTA and RCRA regulations, but some have additional restrictions. Most of the additional state restrictions pertain to routing restrictions, notification, public right-to-know laws, curfews, and size and weight restrictions. Five states have established specific routing requirements for RCRA hazardous wastes—Colorado, New Jersey, Idaho, Kansas, and Rhode Island. New Jersey is the only state that has requirements specific to the transport of CWM.

It should be noted that DOT regulations provide an exemption from portions of its packaging regulations for hazardous materials offered for transport "by, for or to DOD" (49 CFR 173.7(a)) as long as the packaging is in compliance with DOD AR 700-143. DOD also has established its own set of hazardous material regulations that are considered by DOD to provide a level of safety equal to or greater than DOT regulations. One set of DOD regulations is especially pertinent here—DOD TM 38-250, which provides guidance and procedures for preparing hazardous materials by military aircraft and DOD AR 95-27 for transporting hazardous materials by military aircraft.

DOT regulations require that all hazardous materials be properly packaged, marked, labeled, and placarded depending on the type and quantity of material. Most of these requirements flow from the assigned shipping name for the material. The shipper assigns the shipping name and hazard classification after tests have been conducted on the material. The Army believes the recovered CWM would fall under one of two shipping names: (1) waste ammunition, toxic, nonexplosive, without burster or expelling charge, nonfuzed (ID Number UN2016), and (2) waste ammunition, toxic with burster, expelling charge, or propelling charge (ID Numbers UN0020 and UN0021). Based upon this classification, 49 CFR provides detailed instruction as to packaging and other rules that will apply to each shipment of CWM.

Transporting Treated Non-Stockpile Items

The Army's intent is to treat NSCWM on-site using mobile treatment systems such as RRS, SCANS, and EDS and then to transport the resulting waste to a TSDF, including a stockpile CDF. The exception to this would be the NSCWM found on-site at Pine Bluff and Aberdeen. These NSCWM would be treated at the PBNSF and MAPS facilities respectively, thus not requiring off-site transportation.

The waste generated by the mobile treatment systems will still include a substantial volume of waste that may be regulated as hazardous, including neutralent wastes, other industrial chemicals, and various liquid and solid wastes. The Army has conducted bench-scale tests to determine the likely constituents of RRS, SCANS, and EDS waste streams. The composition of the neutralent wastes may include a number of chemicals that are regulated for transportation as both RCRA hazardous waste and DOT hazardous materials and thus will be subject to applicable transportation regulations. The tests show, however, no indication that any of the neutralent waste constituents could not be transported as routine hazardous materials and/or hazardous wastes under existing laws and regulations as long as they are properly packaged, marked, manifested, and shipped.

Thus, based on the information provided by the Army, the transportation of neutralent wastes from the RRS, SCANS, and EDS should not be subject to any restriction beyond the applicable EPA and DOT and associated state regulatory requirements. However, one potential issue that could arise is the public perception related to the residual chemical agents from the EDS waste streams. Although these will be in extremely small amounts, there is the potential for the public to continue to view the overall neutralent waste mixture as tainted with chemical agent and, therefore, of special concern.

TRANSPORTING MOBILE TREATMENT SYSTEMS TO NON-STOCKPILE CWM LOCATIONS

The Army plans to transport the mobile systems, including RRS, EDS, and SCANS, to the recovered CWM site using the most appropriate modes, including by air.

Truck transport is likely to be used for transporting the system to and from air, rail, and water terminals, as needed. System components will be enclosed in trailers and placed in standard truck trailers. Special permits may be required for some components that result in overlength or overweight trucks. The Army will conduct route-specific analyses to ensure that roadbeds, bridges, tunnels, and other infrastructure are suitable for the truck configurations to be used.

The RRS system is expected to require a total of six tractor-trailer combinations to deliver the entire system to the non-stockpile CWM site. The EDS system is expected to require three tractor-trailer combinations. The time for deployment of these systems could range from several days to several weeks, depending on the location of the mobile system at the time it is dispatched and the location of the recovered CWM.

While the overall logistics of these systems is certainly not unusual, the Army will probably have to conduct careful scheduling to ensure that the movement of these systems does not result in disruption to the traveling public. This is especially true of the RRS, where a six-tractor-trailer "convoy effect" could create traffic problems in certain geographic locations (mountain passes, tunnels, winding roads, etc.) and during peak travel times (rush hour in urban/suburban areas).

The Army says it plans to implement mitigation measures to reduce potential impacts. This may include avoiding certain congested roadways, rush hours, and even compensating for improvements to local traffic safety measures and road improvements.

REFERENCE

U.S. Army. 2001. Final Programmatic Environmental Impact Statement. Vol. 1. Aberdeen Proving Ground, Md.: Program Manager for Chemical Demilitarization.

Index

Aberdeen Proving Ground (Maryland), 9, 14, 29, 41, 44, 49, 57, 65, 69, 80-1, 86, 98, *see also* Chemical Transfer Facility, MAPS
ACWA (Assembled Chemical Weapons Assessment (Program)), 5, 21, 37, 40, 64-5, 67-8
Anniston Army Depot (Alabama), 9, 22, 29, 82
   CDF, 9, 82
Assembled Chemical Weapons Assessment (Program). *See* ACWA
binary chemical warfare materiel, 2, 4, 10-11, 14, 19, 21, 24, 35, 41, 49, 50, 53, 88
biotreatment, 27, 38, 40, 81
Blue Grass Army Depot (Kentucky), 9, 22-3, 29, 49, 82
CAIS (chemical agent identification sets), 4-5, 7-8, 10-15, 19, 24, 27, 30-3, 39, 44, 46, 48, 51-2, 58, 61-2, 88, 91-2, 99, *see also* SCANS
CAMDS (Chemical Agent Munitions Disposal System), 4, 15, 18, 24, 49, 53, 83, *see also* Deseret Chemical Depot
CERCLA (Comprehensive Environmental Response, Compensation and Liability Act), 6, 29, 30, 56-8, 61-2, 92, 98, 100-1
CG. *See* phosgene
chemical oxidation, 17-18, 37-8, 40, 48, 51, 54, 61
Chemical Transfer Facility, 4, 15, 18, 19, 24, 53, *see also* Aberdeen Proving Ground
Chemical Weapons Convention, 8, 10, 14, 15, 56, 98
Comprehensive Environmental Response, Compensation and Liability Act. *See* CERCLA
CTF. *See* Chemical Transfer Facility
CWC. *See* Chemical Weapons Convention

DBC. *See* Donovan blast chamber
Deseret Chemical Depot (Utah), 4, 6, 9, 15, 24, 29-31, 49, 54, 57, 82-3, 91, 100-1, *see also* CAMDS
DF, 4-5, 10-11, 14, 19, 21, 23, 35, 41-2, 49, 50, 53
Donovan blast chamber, 2, 6, 18, 27, 32-4, 47-8, 51-2, 54, 65
drill-through valve, 6, 47, 51-2, 54
Dugway Proving Ground (Utah), 14, 26, 49, 52, 56, 98, 100
EDS (Explosive Destruction System), 2, 4-6, 15, 18-19, 25, 28-9, 34, 39, 40, 47-51, 53-4, 57, 61, 65, 68, 87, 92-4, 104-5
electrochemical oxidation, 5, 35, 40, 51
Explosive Destruction System. *See* EDS
gas-phase chemical reduction, 5, 35, 40, 51
GB (sarin), 10-15, 19, 22-4, 26, 28-9, 34-5, 37, 39-41, 43, 56-7, 93, 101
GD (soman), 11-13, 37
H, 11-12, 19, 35, 39
HD, 10-13, 15, 19, 40-1, 91
Johnston Atoll, 1, 9, 22
lewisite, 12, 13, 15, 19, 24, 29, 35, 49
MAPS (Munitions Assessment and Processing System), 2, 6, 15, 18-21, 23-4, 49, 54, 57, 59, 62, 65, 86-7, 100, *see also* Aberdeen Proving Ground
MMD. *See* Munitions Management Device
multiple-round container, 2, 44-5
Munitions Assessment and Processing System. *See* MAPS
Munitions Management Device, 19, 56-7, 101
mustard agent, 10, 15, 19, 22-4, 29, 33, 37, 57, 87, 91, 93, *see also* H, HD

National Research Council, 2, 3, 5, 7, 9, 14, 19, 21, 25, 27-9, 37, 39, 61, 64
nerve agents. *See* GB, GD, VX
neutralization, 4, 5, 9, 15, 17-9, 21, 23, 25, 40-1, 47, 50, 62
Newport Chemical Depot (Indiana), 9, 29
non-stockpile, 5, 8-9, 14, 17, 21, 59
NRC. *See* National Research Council
OB/OD (open burning/open detonation), 18, 42-3, 47, 53-4
persulfate oxidation, 35, 37
Pine Bluff Arsenal (Arkansas), 4-5, 9, 14, 22-3, 29, 35, 49, 51, 53-4, 57, 65, 68-9, 83-5, 87, 92, 98
    PBNSF, 2, 5-6, 15, 18-9, 21-2, 42, 49-51, 58, 62, 86-9, 100-1
phosgene, 10, 13, 15, 19, 24, 35, 56, 91, 93
plasma arc, 4, 18, 21, 34, 40, 47-8, 50-1, 53-4, 65, 87
Pueblo Chemical Depot (Colorado), 9, 22-3, 29, 85
QL, 5, 10-11, 14, 19, 21, 23, 35, 41-2, 49, 53
Rapid Response System. *See* RRS
RCRA. *See* Resource Conservation and Recovery Act
recovered chemical warfare materiel, 7-8, 14, 55, 58, 62, 87
Resource Conservation and Recovery Act, 6, 17, 21, 56, 62, 96-101, 104

Rocky Mountain Arsenal (Colorado), 4, 6, 13, 26, 28, 34, 41, 49, 52, 54, 56-7, 68
RRS (Rapid Response System), 2, 4-5, 15, 18, 27, 29-32, 37, 47, 50-1, 57, 59, 61, 88-9, 91-2, 104-5
sarin. *See* GB
SCANS (Single CAIS Accessing and Neutralization system), 2, 4, 18, 27, 30, 32, 40, 47-8, 52, 92, 104-5
SCWO (supercritical water oxidation), batch mode, 18, 35, 39, 40, 47, 51, 87
continuous mode, 5, 21, 35, 40, 51
solvated electron technology, 5, 40, 51
stockpile, 1-2, 4, 8-10, 15, 17-18, 22-4, 64-5
tent-and-foam, 5, 18, 43, 47, 50, 54
Tooele Chemical Disposal Facility (Utah), 22, 83
treatment, storage, and disposal facility (TSDF), 5, 17-19, 22-5, 41, 51, 58, 61, 100-1
TSDF. *See* treatment, storage, and disposal facility
Umatilla Chemical Depot (Oregon), 9, 22-3, 29, 85
UV oxidation, 35, 40, 51
VX, 10-12, 14-15, 19, 22-3, 37, 40-1
wet air oxidation, 18, 35, 38, 40, 48, 51, 54